STUDIES OF TROPICAL AMERICAN BIRDS

PUBLICATIONS OF THE NUTTALL ORNITHOLOGICAL CLUB, NO. 10
Editor, Raymond A. Paynter, Jr.

STUDIES OF TROPICAL AMERICAN BIRDS

ALEXANDER F. SKUTCH

Dedicated to
DOROTHY LANKESTER
a friend of Costa Rican birds

CAMBRIDGE, MASSACHUSETTS
Published by the Club
1972

Publications of the Nuttall Ornithological Club

1. The Comparative Biology of the Meadowlarks (*Sturnella*) in Wisconsin. Wesley E. Lanyon, 1957. 67 pp., 31 pls.
2. Comparative Breeding Behavior of Four Species of North American Herons. Andrew J. Meyerriecks, 1960. 158 pp., 15 pls.
3. Structural Adaptations of the Head and Neck in the Black Skimmer, *Rynchops nigra* Linnaeus. Richard L. Zusi, 1962. 101 pp., 44 figs.
4. The Rough-winged Swallow *Stelgidopteryx ruficollis* (Vieillot). A Study Based on Its Breeding Biology in Michigan. William A. Lunk, 1962. 155 pp., 19 figs., 3 pls.
5. Ecology and Bioenergetics of the Long-billed Marsh Wren, *Telmatodytes palustris griseus* (Brewster), in Georgia Salt Marshes. Herbert W. Kale II, 1965. 142 pp., 61 tables, 22 figs.
6. Communication and Relationships in the Genus *Tyrannus*. W. John Smith, 1966. 250 pp., 3 tables, 51 figs.
7. Life Histories of Central American Highland Birds. Alexander F. Skutch, 1967. 213 pp., 1 table, 6 figs.
8. Breeding Biology of the Blue-faced Booby, *Sula dactylatra personata*, on Green Island, Kure Atoll, Cameron B. Kepler, 1969. 97 pp., 42 tables, 32 figs.
9. Species Taxa of North American Birds. A Contribution to Comparative Systematics. Ernst Mayr and Lester L. Short, 1970. 127 pp., 8 tables.
10. Studies of Tropical American Birds. Alexander F. Skutch, 1972. 228 pp., 2 tables, 15 figs.

Printed in the United States of America

Publications of the Nuttall Ornithological Club may be obtained from the Nuttall Ornithological Club, c/o Museum of Comparative Zoology, Harvard University, Cambridge, Massachusetts 02138

CONTENTS

INTRODUCTION

The studies included in this book have diverse histories. Some, such as that of the Ringed Kingfisher, were made long ago and withheld from publication in the hope of adding to them—a hope that was never fulfilled. Others, including those of the Pauraque and the Tawny-bellied Euphonia, have grown slowly over the years. Still others, among which are those of the White-fronted Nunbird, the Black-faced Grosbeak, and the Bronzy Hermit, were made rather recently.

As I pass in retrospect more than forty years devoted to the study of tropical birds, I am impressed by the uniqueness of the opportunities for learning about them offered by each season and each locality. In the same locality, one may next year search in vain for a nest of a certain species that he finds this year. Similarly, he may fail to find the same bird in another spot only a few miles distant. Regarding every new kind of nest that I discovered as a unique opportunity to increase our knowledge of Neotropical birds, I have tried to learn something about the habits of its attendants, even when, with other studies in progress, I could not devote much time to it. Indeed, without some sustained watching, one cannot always be certain of the identification of an unfamiliar nest; the literature of tropical American ornithology contains too many misidentifications made by collectors who, apparently, attributed a nest to the first bird that they could shoot near it. In the present state of our knowledge of the birds of tropical America, it seems more important to delineate the broad outlines of the life histories of a wide range of species than to undertake the minutely detailed, statistical studies of single species that are now made in other parts of the world. Such investigations represent a more advanced stage in the study of an avifauna, appropriate for regions where many professional and amateur ornithologists have laid a broad foundation of knowledge.

Because of the uniqueness of the opportunities for learning about the habits, especially the nesting habits, of a large proportion of the tropical birds, I have considered it proper to publish anything significant that I have discovered about them, even if, in consequence of the loss or inaccessibility of nests, or for other reasons, I could not follow all the stages of a nesting. Although observations made in a laboratory can commonly be repeated and substantiated, this is far from true of those made in the field with animals so elusive as birds. Doubtless every bird-watcher has recorded observations that he has never been able to duplicate, and which others may not for many years confirm.

About half of these studies were made in the Valley of El General, at the head of the Río Térraba drainage on the Pacific side of southern

Costa Rica, and especially on our farm, "Los Cusingos," situated at about 2,400 feet above sea level beside the Río Peña Blanca, an affluent of the Río Térraba. A number of others were made at "La Selva," on the left bank of the Río Puerto Viejo just above its confluence with the Río Sarapiquí, which in turn is a tributary of the Río San Juan in the Caribbean lowlands of northern Costa Rica. These two localities are described in some detail in a recent book (Skutch, 1971b.) Some of my observations on the Pauraque and Massena Trogon were made long ago in the Lancetilla Valley, just inland from Tela on the Caribbean coast of Honduras. I studied nests of the Ringed Kingfisher along the Río Morjá, which flows into the Río Motagua near Quiriguá, Guatemala. The Unicolored Tapaculo and the White-capped Dipper were found nesting near Baños, Province of Tungurahua, on the eastern slope of the Ecuadorian Andes.

Our work at La Selva in 1967 and 1968 was supported by a grant from the Frank M. Chapman Memorial Fund of the American Museum of Natural History. In 1967, Dr. and Mrs. Leslie R. Holdridge, then proprietors of La Selva, kindly permitted us to occupy their house and helped us in many ways. In the following year, the Organization for Tropical Studies, which meanwhile had acquired the property, extended the same courtesies through its resident director, Señor Jorge Campabadal. My wife, Pamela, and son, Edwin, helped in the field, the latter alternating with me in long-continued watches at nests. To all these individuals and institutions, I am most grateful.

The manuscript was accepted by the Nuttall Ornithological Club 20 September 1971.

ALEXANDER F. SKUTCH

Finca "Los Cusingos"
San Isidro del General
Costa Rica

Family CAPRIMULGIDAE

PAURAQUE

Nyctidromus albicollis

The Pauraque is a long-tailed, terrestrial goatsucker or nightjar about 10 inches long. The browns, buffs, grays, and black of its soft plumage form an intricate pattern of delicately blended shades, lovely to contemplate at close range and at leisure, but so tedious to describe that the reader who desires more details of the bird's appearance must be referred to works of descriptive ornithology, such as Ridgway (1914:537-539), or better, the illustrated guides to North American birds. Recognition marks are the broad band of white on each wing, the largely white outer tail feathers, and the white throat of the male. On the slightly smaller female the white areas are much reduced in size and more or less tinged with ochraceous buff. Pauraques exhibit two distinct color-phases, one gray-brown and the other tawny, but the pattern is the same in both phases. The small bill and inconspicuous feet are brown, the large eyes dark brown.

One of the most common and widespread birds of tropical America, the Pauraque has an enormous range from the valley of the Río Grande in southern Texas to Paraguay and northern Argentina, and from Ecuador and Peru to Brazil, the Guianas, and Trinidad. Most abundant in the warm lowlands, it has been recorded as high as 5,700 feet in Guatemala (Land, 1962:273) and 6,500 feet in the western Andes of Colombia, where evidently it was moderately abundant and breeding (Miller, 1963:10). The report of Carriker (1910:502) that in Costa Rica the Pauraque was present "up to 9,000 and even 10,000 feet on the Volcanoes de Irazú and Turrialba," and was found nesting at nearly 9,000 feet on the former, is puzzling and requires explanation. The highest point at which I have ever met this bird in Central America was 5,500 feet, at Vara Blanca on the northern slope of Costa Rica's Cordillera Central, where in the course of a year I saw a single individual, evidently a straggler from lower altitudes. During five months on the southern slope of the Barba massif, where conditions are much the same as on neighboring Volcán Irazú, I failed to see a Pauraque even as high as 6,500 feet.

Like many other birds of very wide geographical range, the Pauraque exhibits great ecological tolerance. It is found in semiarid regions with light vegetation as well as in clearings amid the heavy rainforest of some of the wettest districts of tropical America. Griscom (1932:190) pointed out how sensitively the shade of the Pauraque's plumage responds to the humidity of its environment, being darker in the wetter districts. So pronounced is this variation that the bird's coloration might be used as an indication of the local rainfall!

Through most of its range the Pauraque resides the year around in the area where it breeds, but the northernmost race, *N. albicollis merrilli,* is to some extent migratory. Although many individuals of this race winter throughout their breeding range in extreme southeastern Texas and adjacent portions of northeastern Mexico, others migrate southward to Veracruz and Puebla to pass the winter months (Friedmann *et al.*, 1950:154).

Avoiding heavy forest, the Pauraque inhabits all sorts of semiopen and cultivated country, including swampy areas. Its optimum habitat contains woodland edge, thickets, canefields, or other low, dense vegetation where the birds can rest undisturbed by day, alternating with open spaces where in the twilight and on moonlit nights they can pursue their aerial insect-catching with little impediment. Often the open space where they forage is a lawn, a pasture, a cultivated field, a road or railroad, especially where these ways of communication are bordered by bushy growth.

All through the day the Pauraque drowses quietly in the shade of some second-growth thicket near the edge of a clearing, or in a plantation of coffee or bananas, on brown dead leaves with which his coat of mottled black, browns, and buffs blends so marvelously well that anyone penetrating his retreat might step upon him without seeing him. Fortunately for the bird, he is not sound asleep, and in time to save himself from being trampled he starts up in front of you with alarming suddenness, as though the ground litter itself had come to life. Rudely aroused from his repose, he is as likely to dart toward as away from the intruder; he circles around in silent flight, revealing on his spread wings and tail conspicuous areas of white, which vanish as he settles on the ground and folds his limbs, often not far from his original position and in full view. Rarely when disturbed he alights on a low branch, on which he rests with his body parallel to the limb.

In the evening, as other birds retire to their roosts or dormitories in the waning light, the crepuscular Pauraques come forth into the open to feed and to sing. Resting on open ground, a rock, or the flat top of a fence post and tirelessly repeating their song, they continually raise themselves lightly on their feet without severing contact with the supporting surface. They make little upward jumps to catch passing insects, falling back into the same spot, and short circuitous flights which often terminate at or quite near their starting point, catching insects in their capacious mouths as they go. They consume a great variety of night-flying insects, including moths and beetles of many kinds; sometimes they concentrate on fireflies (Bent, 1940:203-204).

When one watches Pauraques in the dusk, and now and again catches the glint of their large, oval eyes as they are turned up, down, and sideways, looking for small volitant creatures, the effect is weird. In the beam of a flashlight, or the headlight of a passing car, their eyes return a ruby reflection. On rough country roads, where cars

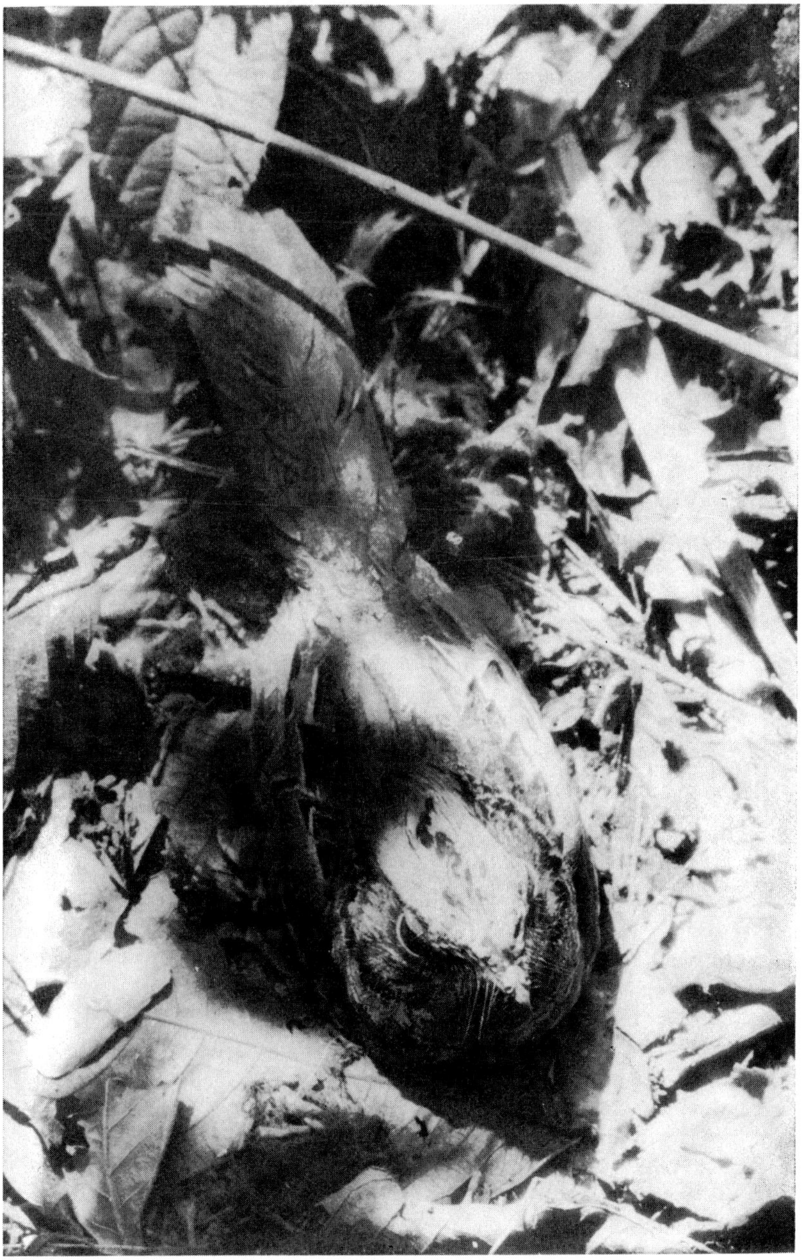

FIG. 1. Female Pauraque incubating. Lancetilla Valley, Honduras, 15 May 1930.

cannot go very fast, the Pauraques that have been calling and foraging in the roadway often fly ahead of the approaching vehicle, alight again in the road, fly once more as the car bears down on them, finally to veer aside into the bordering woods or thickets and permit the car to pass. Years ago, when we travelled by night through the extensive grassy swamp lands near the Sixaola River in northwestern Panama, the headlights of our motor tramcar frequently revealed a Pauraque's twin ruby lights between the rails. The birds would delay until the rushing car was almost upon them before flitting deftly aside.

Voice

As night approaches and diurnal birds cease to sing, the Pauraque breaks his daylong silence. *Kw kw* he calls, in low tones audible only when he is close by, as though still not fully awake. Only after several repetitions of these preliminary notes does the full song burst forth: *kw kw, kw kw, kw kw, kw-ah-réo.* After the sunset glow has faded and the Pauraques have warmed up to their chorus, answering each other from various parts of the clearing, they seldom use the preliminary notes but utter directly their full *kw-ah-réo,* loud, clear, ringing, and just a trifle tremulous. This call is not so insistent and wearing on the nerves as the incessantly repeated admonition of the North American Whip-poor-will, but softer and more melodious, even somewhat plaintive, which the cry of the northern bird certainly is not. The song of *Nyctidromus* is responsible for a number of its vernacular names, including Pauraque itself and Cuyéo, as this goatsucker is widely called in Costa Rica. On clear, moonlit nights the Pauraques repeat this song at short intervals for hours together; but on moonless or darkly clouded nights they become comparatively silent after singing for a short while in the dusk, to resume their full chorus for an equally brief period in the earliest dawn. Then, as Chapman (1929:38-39) put it in his inimitable way: "At about 6 o'clock, with a *hip-hip; hip-hip; hip-hip-hooray,* the Parauque Goatsucker announces the coming of day."

In the Valley of El General, the Pauraques, who have been silent for months, begin to sing as the long wet season approaches its end. I have heard one sing in the moonlight after an exceptionally dry week at the end of October. Song increases, but is still sporadic, during November and December, to reach its climax on clear, moonlit nights in the dry months of January, February, and March. The rains return in late March or April and in May usually become heavy, bringing to an end the Pauraque's season of song. By the middle of a wet May, even a bright moon fails to stimulate the birds to sing. On the Caribbean coast of Honduras, where the drier months come later in the year, I heard Pauraques sing freely for as much as a quarter- or even half-hour at nightfall in late September. In rainy October, they soon fell silent.

In addition to their clear, far-carrying song, Pauraques utter a variety of low, guttural and croaking notes, and a high, clear, soprano *whoo*. They also have a special call, described beyond, to guide their chicks to themselves.

Eggs

Since the Pauraque builds no nest but lays its eggs directly upon the ground, or on such litter as it finds there, heavy rains are detrimental to them and to the semialtricial young. Hence the Pauraque breeds in the drier part of the year, which in the northern tropics of America nearly everywhere falls within the first four or five months. My earliest date for eggs is 16 February 1936, when I found a completed set in the Valley of El General at an altitude of nearly 3,000 feet. For this region I have only one other record of a brood started in February. In the Motagua Valley of Guatemala, near Quiriguá, a set was laid at the very end of February 1932. In El General, laying is at its height in March, in which month 11 sets have been recorded, and continues into April, when five sets were recorded. In this region I have no record of laying in May, although some of the eggs found in late April would have hatched in this month. On the north coast of Honduras, however, we noticed one set in May and three in June.

In Nicaragua, as in Costa Rica, breeding takes place between February or March and May (Richmond, 1893; Huber, 1932). At the northern limit of the Pauraque's range, in Texas, 66 records of eggs fall between 5 March and 26 June, half of them between 15 April and 16 May (Bent, 1940:205). Elsewhere in the northern tropics, the situation is similar: in Trinidad the Pauraque breeds from February to July, with the peak of the season in the dry months of February, March, and April (Belcher and Smooker, 1936:23). In northeastern Venezuela nesting occurs from March to June (Friedmann and Smith, 1955:520). These and a number of other scattered egg dates that need not be listed here give a consistent picture of breeding, north of about latitude 5° N, between February and June, beginning a few weeks earlier in the southern part of this region, as in Central America and Trinidad, than in northern Mexico and southern Texas. The only mention I find of late breeding is for the coastal range of Venezuela, where at Rancho Grande the season is given as May to August, without any indication of the evidence on which this statement is based (Schäfer and Phelps, 1954:67).

The Pauraque's eggs are often laid in spots somewhat more open than those in which these birds rest by day. Frequently they are situated in a little opening in the midst of a thicket. In pastures and fields they usually lie in the shade of an isolated shrub or small tree, and in banana or coffee plantations thay are deposited near the base of one of these plants. In Honduras a set was found in a little pocket of pebbly sand shaded by low shrubs, between stones and boulders on

FIG. 2. Eggs of Pauraque. Lancetilla Valley, Honduras, 20 May 1930.

the barren flood plain of a river, so near the stream that they were washed away by the next high water. Another set was laid in an open space amid the dry bushes and other vegetation of a field cut over for burning, where they barely escaped the fire. Usually the eggs rest upon brown fallen leaves and other litter, but occasionally they are in contact with bare ground, sand, or pebbles. There is never any indication of nest building.

The set regularly consists of two eggs, apparently never more. When, as rarely happens, a single egg is incubated, its companion was probably lost. I have no information on the hour when the first egg of a set is deposited, but the second is laid rather late in the afternoon. Thus, at 17:30 on 27 February 1932, I found a male Pauraque covering a single egg. The second was laid between 15:40 and 18:15 on 29 February. In the two-day interval between the laying of the eggs, the first was found to be covered four times by the male and once by the female, on five visits of inspection. At 08:45 on 15 April 1946 I again discovered a male sitting upon a single egg. At 14:05 on the same day he was also covering the single egg, but at 17:40 the female was incubating two eggs. In both instances the egg that completed the set was laid long before the Pauraques became active in the evening twilight. It is almost as though a diurnal bird were to lay its eggs in the darkness before dawn. Weller (1958:56) reported that a Common Nighthawk laid its second egg late in the morning, between 10:30 and 11:30.

There is often little difference in shape between the two ends of the eggs, both being almost equally blunt. The shell is only slightly glossy. In color the eggs are light buff or pinkish buff, spotted and blotched with shades of brown and lilac. The marking may be light or heavy, but usually it is fairly uniform over the whole surface. The measurements of 12 eggs in Central America average 29.8 by 22.0 mm. Those showing the four extremes measured *31.8* by *20.2*, 30.2 by *23.8*, and *27.4* by 21.4 mm.

Lying on a bed of brown leaves, the Pauraque's eggs, much lighter in color, are far easier to detect than the bird itself. But since, from the time the first is laid, they are almost constantly covered by the parents, the lack of closer assimilation to their background is probably no great detriment to their safety.

Incubation

Both sexes incubate. At the "nests" that I studied in Honduras and Guatemala in 1930 and 1932, I did not watch from a blind during incubation, but on a number of visits of inspection I recorded the sex of the parent covering the eggs. At the first nest 40 daytime visits revealed the male incubating 24 times, the female 16 times. Taking all my records for these years together, on 62 visits to nests I found the male sitting 39 times and the female 23 times. Thus the male performed the larger share of daytime incubation, but the female took a substantial part in this duty. Both took turns on the eggs in the morning as well as in the afternoon. At the first nest they seemed to be replacing each other at intervals of two or three hours, but the sessions were not accurately timed. On a few nocturnal visits, only the female was present. Once, when driven from her eggs on a moonless night, she returned to them within 25 minutes.

My most careful studies of incubation were made at two nests at Los Cusingos in 1962, especially at one in the pasture on the hilltop behind the house. Here the male did nearly all the diurnal incubation. On 32 of 36 visits of inspection he was covering the eggs. The female was present only once, and on three occasions the eggs were unattended. In addition to these brief visits of inspection I spent about 32 hours watching from a blind. On 23 and again on 31 March, I watched the male Pauraque sit continuously for six hours of the forenoon. When I returned at 13:43 on the latter date he was still on the eggs, where he remained for the next 4¾ hours, or until nightfall. Evidently, if undisturbed the male incubated continuously throughout the day. Watching from concealment in full daylight, I saw the female incubate only once. During my vigil on the afternoon of 29 March a strong gust of wind shook the blind and frightened the male from the eggs.

They remained exposed for nearly an hour, then the female came and covered them for the remainder of the afternoon, or for 3½ hours continuously.

As daylight waned the male Pauraque flew from the eggs at hours ranging from 18:11 to 18:25. The eggs then remained exposed until it became too dark to see the bird arrive to take charge of them, so I left the blind and made brief visits of inspection with a flashlight. Usually within 20 to 35 minutes after the male's departure I would find the female incubating. Exceptional behavior was witnessed on the evening of 27 March, when at 18:17 the female alighted on the ground near her incubating mate, then flew away. At 18:21, he left. He was still absent at 18:35 but had resumed incubation at 18:45. At 19:00, the female was covering the eggs.

On 29 March, when the female took a long afternoon session on the eggs after her mate had been frightened away, she sat until evening. At 18:15, when the light was failing, the male at last emerged from the neighboring woods into which he had retreated. When he appeared his incubating mate uttered a single low, soprano note. After flying northward along the woodland's edge he returned at 18:18 and alighted about one foot from the female. They uttered low, harsh notes. Then the male rose up and alighted on the female's rump. She delayed to leave, possibly because a Blue-diademed Motmot was perching directly above them. Soon she crawled out from beneath her mate and flew away. Thereupon he waddled onto the eggs and adjusted them with his bill. He covered them for only seven minutes. At 18:25, his usual time for leaving, he flew northward in the dusk. In the distance I heard a single song.

A few visits with a flashlight up to midnight always revealed the female incubating. When I returned by the light of the waning moon, shortly before dawn, I would sometimes find the male on the eggs, sometimes the female. After determining the sex of the bird I turned off the flashlight to watch from the blind, but sometimes I failed to notice the Pauraque's departure in the dim light. It was clear, however, that the course of events varied from dawn to dawn. On 23 March, for example, when the nearly full moon was shining brightly, I arrived at 05:15, just as dawn was touching the eastern horizon. At 05:22 the male approached, calling *kw kw, kw kw, kw kw, kw-ah-réo,* and alighted beside the nest. Thereupon the female left. With an airy bounce he covered the short distance that separated him from the eggs and sat upon them for the rest of the morning.

On 28 March, when a half moon was shining at the zenith, I found the female sitting at 05:00. The nest was in the shadow of an overhanging branch and I failed to see her departure, but at 05:27 she returned. Two minutes later her mate alighted beside her, she flew away, and he rose lightly to come down on the eggs. Even to cover a short distance he preferred to fly or bounce rather than to walk. On

27 March when I found the male sitting at 04:30, he left at 05:11, flashing the white of his tail. He was absent for 19 minutes, returned at 05:30, sat for six minutes, left at 05:36, and after remaining away for three minutes returned at 05:39 to begin his long diurnal session. On 31 March he arrived at 05:28, incubated for 11 minutes, was absent for four minutes, then at 05:43 returned to cover the eggs for the remainder of the morning. On moonless nights the short intervals in the evening and morning twilight when both parents are absent from the eggs are probably when they chiefly forage.

The eggs disappeared from this nest when they were on the point of hatching. In a coffee plantation 1,000 feet away I found on 19 April a single egg, and by the following morning the second egg lay beside it. The male covered these eggs continuously during five hours of the morning of 25 April, when I watched from a blind. He was also incubating when I resumed my vigil at 12:20 next day. He sat continuously until 15:17 when, for no apparent reason, he flew from the eggs to alight among low weeds about three yards away. Earlier in the afternoon he had been drenched by a hard shower. Now he preened, shook himself vigorously, and swayed from side to side, then drowsed with nearly closed eyes. After an interval of repose these activities, or some of them, were resumed. Then followed another period of motionless rest. As the sun sank lower these intervals of inactivity grew longer, while the preening, shaking, and swaying were more briefly performed. After he had been a full 2½ hours at this same spot, always facing his neglected eggs, I left him there. His mate had not appeared all afternoon. On eight visits of inspection in the morning and afternoon it was always the male whom I found incubating here. I did not visit this nest by night.

At a nest with a single egg found in April of 1965, the male was incubating on 12 of 18 daytime visits, the female on three, and the egg was unattended on three visits. It is evident that there is considerable variation, which may be to some extent geographical as well as individual, in the part that the female takes in diurnal incubation, but that at most, if not all, nests the male is present by day much more than the female. Because of difficulties of observation we have little information on nocturnal incubation, which seems to be performed chiefly by the female. It would not be surprising if the pattern of nocturnal incubation varied with the stage of the moon, but it might be difficult to demonstrate this, for in the shadows the Pauraques, especially females with little white on wings and tail, are hard to see even on moonlit nights.

A Pauraque coming to incubate carefully adjusts the eggs beneath itself with its bill, then moves its body slightly from side to side as it snuggles down on them. In the midst of a session of incubation it rarely touches the eggs with its bill, but from time to time adjusts them with slight sideward, and sometimes also vertical, movements of its body, much as a broody domestic hen does. Once, after a drenching shower, a male slid sideways over his eggs four times in 3/4 of an hour; and on a warm afternoon another male did so 10 times in 1½ hours; but usually such adjustments are made at longer intervals.

The Pauraque's eyes are directed sideward. When viewed from behind he seems to be looking squarely at his observer. A patch of dark feathers in the center of the male's hindhead simulates a bill, and it is easy to imagine that the bird has turned his head straight backward to look at you. I noticed this weird effect while watching from the blind. The pattern of black markings on an American Sparrow Hawk's head likewise creates the illusion that he is looking straight at you when you view him from behind. Possibly there is some protective advantage in this visual deception.

When the Pauraque is at rest with the white patches on his wings and tail concealed and his mottled plumage blending with the ground litter, perhaps his most revealing feature is his large, lustrous, dark eyes. While incubating by day he rarely opens them fully, and he rarely keeps them closed for more than a few seconds together—half a minute was the longest doze that I noticed in many hours of watching. His eyelids seldom continue long in one position but are in almost constant slow movement, changing the aperture of the eye. Now his eye closes except for a narrow gap over the posterior half; now the lids are everywhere tightly pressed together; now they open with a narrow slit that extends horizontally across their whole width; now they slowly draw together again. Rarely do the lids open as much as halfway. Both of them move up and down. I have watched the same restless movements of the eyelids of a Common Potoo incubating or brooding by day. Even with its eyes nearly closed, as though dozing, the Pauraque sees very well what is happening around it.

The nest on the hilltop was exposed to full sunshine for only a few hours in the early morning, and the incubating male always sat with his tail toward the sun, as incubating female Common Nighthawks do (Weller, 1958). Indeed, he maintained this orientation, facing the nearby woods, throughout the day. When full sunshine falls upon an incubating Pauraque, he sits higher than usual. His throat flutters, too rapidly to count the vibrations, while his bill is kept slightly open. The movement of the white throat catches the eye, which finds it difficult to detect the remainder of his body. To derive full advantage of their protective coloration, Pauraques should nest in the shade, where they have less need to cool themselves by the gular flutter.

Early one afternoon an incubating male opened and closed his bill with a slight, rapid movement that produced a low, clicking sound and little flashes of light color from the interior of his mouth. Bouts of rapid bill snapping were repeated at intervals for an hour. Most of the time, however, this male did not engage in this puzzling activity.

Rarely an incubating Pauraque yawns widely, exposing the flesh-colored interior of his cavernous mouth.

In a hard shower an incubating male sat higher than usual, with his head drawn back and the white patch on his throat showing prominently. He kept his eyes opened more widely than in dry weather, with half or more of the eyeball exposed. From time to time he blinked or swallowed. When the sun burst forth again and the Gray's Thrushes were singing gloriously, the Pauraque sank down to the normal incubating posture and his eyes narrowed to a slit. His plumage appeared slightly bedraggled from the wetting. Soon after this he rose from the eggs to alight nearby, where he preened, shook himself vigorously, and scratched, as already described. It is understandable why Pauraques were seldom seen to preen, and never vigorosuly, while covering their eggs; such movements would cancel their excellent protective coloration. But why did this bird remain looking toward, rather than warming, his eggs for so long after he had finished arranging his plumage? Apparently the male Pauraque's drive to incubate grows weak as the afternoon advances. At the nest on the hilltop the male sometimes neglected the eggs for hours if for any reason he left them in the afternoon. If driven from them in the morning he returned more promptly.

There is, or was, a fairly widespread belief that if a goatsucker's eggs or young are touched or otherwise disturbed, the parents frequently move them to a safer spot. Aubudon (1831) published a circumstantial account of how a pair of Chuck-will's-widows transported their eggs beyond view, each parent carrying a single egg in its capacious mouth. Long before Ganier (1964) discredited this tale,[1] I was eager to see for myself whether Pauraques ever moved their eggs or young, as some of its relatives had been reported to do. By repeated visits to nests, by measuring eggs, and by setting blinds before them, I gave several pairs of Pauraques ample provocation to move their eggs to a more secluded place, but the parents never shifted them even a few inches from their "nest" as a result of my interference.

The case was quite different when *I* moved the eggs from the spot where the female had laid them. A pair in which I was interested was incubating on a site that had become infested with fire ants, which burrowed in the ground below the eggs. How the Pauraques endured to sit there as long as they did was a mystery to me, but evidently their thick plumage afforded sufficient protection from the stinging ants. I knew that, although the eggs were fairly safe in this position as long as they remained uncracked, the chicks would be destroyed by the ants as soon as they chipped the shells. Accordingly, when I heard tapping inside the eggs I moved them a few inches beyond the ant-infested area,

[1] Ferguson (1967) has reopened the question by presenting circumstantial evidence that a Chuck-will's-widow's eggs were moved 15 feet.

to give them a minimum of safety. The next day I found that they had been shifted back to a few inches on the opposite side of the original spot, where they were still above the ant's nest. Again I moved them to a clean space, this time a foot away, but again the Pauraques brought them back to their chosen spot! By this time one of the chicks, after two days of tapping, had pierced its shell but the ants swarmed in through the perforation and devoured it. I did not see how the Pauraques moved their eggs but doubtless it was by rolling them over the ground with their bills, feet, and breast feathers, as nighthawks and other goatsuckers do. After this the second egg was abandoned.

The incubation period of a bird like the Pauraque is most difficult to determine because, in the complete absence of a nest, one cannot tell in advance where the eggs will be laid. Only if one happens to find the first egg of a set before the second appears is he able to time the laying of the latter. In the three instances when I did this, and so could date the beginning of incubation, the eggs were prematurely destroyed. Probably the incubation period of the Pauraque is about 19 or 20 days, as in other caprimulgids.

Young

The first Pauraques' nest that I ever watched was situated in a low thicket beneath a colony of Montezuma Oropéndolas, in the Lancetilla Valley of northern Honduras. I enjoyed the contrast between the oropéndolas incubating their eggs in carefully woven pouches hanging 60 feet overhead and the Pauraques hatching theirs on the ground below, with no trace of a nest. Despite the great difference in the extent of the preparations which the two kinds of birds made for their nestlings, both were equally good parents. Having provided an exceptionally safe receptacle for their eggs and young, the oropéndolas made little attempt to lure or drive away potential enemies. But the Pauraques, who had placed their eggs on the surface of the ground where they were exposed to the greatest number of perils, did their best to entice away intruders by means of convincing distraction displays. As I cut my way with a machete into the little glade where they nested, the female, who happened to be covering the two eggs, rose from them in her light, airy way and fluttered down a few paces off, where she grovelled among the leaves, pressing her breast to the ground and beating it with her extended wings, uttering the while low, croaking sounds. I followed and she repeated the same self-abasement several times, between grovellings bouncing ahead of me lightly, until she had led me into the dense bushes beyond the glade. The male appeared on the scene for a minute only and did not second her efforts to lure me away.

At this nest the eggs hatched on successive days. For 27 hours I had heard the rhythmic *tap, tap, tap,* as regular as the ticking of a clock, and occasional peepings from the chicks within, before at length they burst from the shells. The newly hatched Pauraques were nearly covered with the softest down, which was chestnut-brown on the sides of the head and the back. The crown and a stripe along the center of the back were tawny, as were the sides and under parts of the body. The chicks' eyes were partly open, their bills mere bumps extending beyond the prominent nostrils. The parents promptly removed the empty shells from the vicinity. When the mother covered the nestlings, she permitted me to approach to within a few feet before leaving them, to flutter around very close to me, grovelling, beating the ground with her wings, and uttering low, mournful croaks. Even the male now behaved in this fashion but his acting was neither so spirited nor so prolonged.

It rained during the nestlings' first night and when I returned the following morning they had vanished; but their mother alighted almost at my feet and grovelled on the leaves. After much searching I found them in a denser part of the thicket, about two yards from the spot where they had hatched, which was more open and exposed to the rain. But how had they been moved? I recalled reading that various relatives of the Pauraque carried their young, like kittens, in their capacious mouths and I wished to see if this was true. I set up a blind in the glade and placed the downy chicks directly in front of it, about two feet from the point where their father had been brooding them. I had not long to wait in the blind until he returned,

FIG. 3. Chicks of Pauraque, one and two days old. Lancetilla Valley, Honduras, 22 May 1930.

advancing cautiously by a series of short flights more like hops. I watched intently, fully expecting to see him carry off his chicks one by one in his mouth. Instead of this he settled down in the exact spot whence I had removed the little ones and uttered a low call, a sound between the clucking of a domestic hen and the croaking of a frog. The chicks, who at that moment were reposing quietly, suddenly started up, stretched forward their necks, and by a series of short, weak hops advanced over the rough, leaf-strewn surface of the glade, straight toward their father, under whose soft feathers they pressed. They were then between one and two days old.

I also wished to see how and when the young Pauraques were fed. At dusk their mother returned and appeared to give them food, but it was too dark in the glade to distinguish details and the beam of an electric torch frightened her away. When she came again she called the chicks to a more distant spot, where she fed them twice more, then settled down to brood them.

The following morning the young Pauraques attempted to hop away when I came near them. Both parents took turns at brooding them throughout the day, relieving each other at intervals of two or three hours. In the evening I moved the blind to command their new position and waited to see them receive their supper—or, perhaps more properly, breakfast. As the dusk deepened beneath the bushes the mother flew up, fed the chicks, and remained brooding them. Again it was too dark to distinguish details but I forebore to use the light, expecting better results later. After about five minutes the chicks emerged from beneath her and stretched up in front of her and, touching her mouth with their bills, begged for more food. She found a little more in her crop and passed it to them, but still they were not satisfied and persisted in their demands until she pushed them down with her bill. In a few minutes she left once more. In the distance behind me I could hear her not unmusical *whooo, whooo,* uttered in a soprano voice, like a girl on a hillside calling to someone. The chicks answered with loud *peep's* but they did not budge from the spot where she had left them.

Presently the parent returned, fluttering noiselessly over the ground in an indirect course, and began to feed the chicks again. It was quite dark now but at the critical moment I pushed the button of a powerful flashlight. The parent was resting on the ground in her customary position, looking toward me with eyes that shone like rubies in the beam of light. The little ones seemed to stand on tiptoe in front of her, their necks stretched up, bringing their heads on a level with her mouth. One was being fed while the other waited impatiently for its turn. Although the parent's *mouth* was so big that she seemed capable of swallowing the nestling whole, her little *bill* was inserted into its widely open mouth, just as a hummingbird does, and with distended throat and convulsive movements of her body she regurgitated into it the insects she had captured. When she had given enough to the first chick she paused with head erect, clearly much alarmed by the light; but the second chick continued its silent importunity, to which the parent yielded despite her fear, feeding it in the same manner. Then she went off to catch more insects.

Presently one of the parents entered the thicket again, but instead of going to the chicks it settled on the ground behind the blind, fully 20 feet from them, and began the low croaking-clucking call which drew them as a magnet draws a needle. The little Pauraques, now two and three days old, rose up and started hopping toward the voice, peeping as they went through the darkness. Directly in their path was a young banana plant that I had felled while clearing a place for the blind. The big, slippery leaves lay in a tangled mass that loomed above them, a seemingly insurmountable barrier; but with perseverance they pushed onward, lured by the continuing calls. I completely lost sight of them in the obscurity but for many minutes their weak cries emerged from the dark mass. At last they struggled out from the farther side of the banana plant and thenceforth their path, although not free from obstacles, was comparatively easy. Finally they reached a haven beneath their parent's sheltering wings.

I tried to learn whether the chicks were fed in the morning as well as in the evening. Twice in the late afternoon I set the blind in front of them, but each time I returned before dawn only to find that they had moved to a point where I could not watch them from concealment. On the second morning they completely escaped my search and I never found them again.

Parent Pauraques sometimes call away the first chick before the second has hatched. One evening I found a newborn chick lying beside an empty shell and an unhatched egg. On the following morning the chick was being brooded three yards away. The second egg, in which the occupant was tapping and peeping weakly, had been left cold and deserted in the "nest," along with the empty shell. Doubtless it would never have hatched if I had not placed it beside the older chick.

Because of the mobility of even newly hatched Pauraques and the difficulty of observation at night, I have not again succeeded in watching the act of feeding. In an effort to learn more about the feeding and development of the young, I once enclosed two feathered but still flightless Pauraques in a palisade of sticks, open above so that the parents might enter and attend them. On the following day one was lying dead within the enclosure while the other, having managed to escape, was brooded by its father nearby. I did not again try to restrain the movements of young Pauraques.

The parents attend the young until they are well feathered. Occasionally an adult has risen from the ground in front of me, where it had been brooding, or at least resting close beside, half-grown young; since my attention was first drawn to the adult as it left the ground I was usually not certain whether it had been above or near the young. While its father was giving a distraction display, one young Pauraque, well feathered but small enough to fit into the palm of my hand, tried to escape by hopping over the ground with its dark wings raised high above its back, repeating a queer little *whoo*. On another occasion, after an adult male flew up from the ground as I approached, a juvenile arose from the same spot and flew easily about 20 feet, revealing white tips on the outer tail feathers. It alighted on a branch about a foot above the ground, and when I approached it resumed its flight. This juvenile seemed scarcely bigger than my palm. Both parents attend feathered young by day but I have found the male with them more often than the female.

The young Pauraque combines in an interesting manner the characteristics of altricial nestlings and precocial chicks. In its total dependence upon its parents for food during the early days of its life it is typically altricial. In its unbroken garment of soft natal down it is as precocial as any gallinaceous bird or tinamou. In its limited ability to move over the ground from the day it hatches it is intermediate between altricial and precocial chicks.

Distraction Display

Like many other ground-nesting birds, Pauraques have a spectacular distraction display, by which they attempt to lure from their nests, or young, animals that might harm them. Individual Pauraques differ greatly in their proneness to give distraction displays and the intensity of their performances. At the nest beneath the oropéndolas' tree in Honduras the female gave spirited displays from the day I found her, but the male did not display before me until the first egg hatched, ten days later, and even then his performance was only lukewarm. At the other nest to which I gave most attention, that on the hilltop at Los Cusingos, neither parent ever displayed to me during the 15 days while they incubated but both displayed earnestly before an opossum, 13 days before their eggs were pipped. As I examined these eggs when they were on the point of hatching, the male parent flew in circles close around me, uttering low notes, but he gave no distraction display.

Distraction displays are frequently given by parents disturbed while brooding or guarding their young. One of the most vigorous and long-continued displays that I have ever witnessed was given by a female whose two young could fly quite well. While I searched for them at the woodland's edge she again and again alighted near me to grovel and beat her wings on the ground, continuing for a quarter of an hour, until I abandoned the fruitless quest. On the following day, in the same place, she displayed again and this time I stirred up the juveniles she was guarding.

I have already described the Pauraque's distraction display as I first witnessed it beneath the oropéndolas' tree, but on later occasions other details were noticed. In a typical display, the parent lies flat with widely spread wings that are fluttered or beaten against the ground, sometimes rather rapidly, sometimes slowly, as though the bird can hardly move them. Low throaty sounds, often suggestive of agony, accompany the performance. Occasionally the displaying bird hisses with widely opened mouth or it may make a low rattling sound. It may drag itself over the ground or, as the intruder approaches, rise lightly to come down ahead of him and continue to grovel. Sometimes the displaying Pauraque flies toward the intruder, passing close by him to settle on the opposite side and repeat the act, trying to lure him in a different direction.

The "injury-feigning" display before the opossum was one of the few that I have seen a bird perform in front of an animal, other than man. I had entered my blind before the Pauraques' nest by the light of a nearly full moon in an overcast sky. Distant Pauraques and Gray's Thrushes were singing. I could not distinguish the sex of the Pauraque covering the two eggs. While the light was still dim the mate of this bird emerged from the neighboring low woods displaying before the marsupial. It fluttered on the ground in front of the advancing animal, waiting until the latter was about a foot away before rising to flit ahead and grovel on the ground again. The opossum continued to lumber toward the bird at a uniform speed. They advanced toward the blind and presently the Pauraque, fluttering on the ground, almost touched the edge of the cloth. Still pursuing the bird, the opossum nearly bumped into the blind, then turned right and ambled down to the edge of the woods, into which it vanished, having been led in a wide semicircle around the eggs. The other Pauraque had left the nest to join in the display but it hardly influenced the animal's course.

Enemies

Doubtless adult Pauraques have their enemies but I know nothing about them. The eggs and young, lying on the open ground, are exposed to many perils. I have already told how fire ants caused the loss of one set of eggs although I had moved them to a safer spot. On another occasion I found a newly hatched chick dead and swarming with fire ants, while the father brooded its sibling a yard away. Another set of eggs was apparently trampled by cattle, which must be a frequent occurrence since Pauraques so often nest in pastures. Still another set, laid on the stony flood plain of a stream, was washed away by high water.

The ground fires, so frequent in the dry season, undoubtedly destroy many eggs and young. Pauraques begin to breed about the time when woods and thickets have been cut by farmers and are waiting to be burned to clear the land for planting. The Pauraques not infrequently lay their eggs amid the drying vegetation. On two occasions I have known the nests to escape unharmed in a little islet that remained unburned in the midst of the conflagration, but probably they are rarely so fortunate. Once the fire roasted a newly laid egg, which I ate, finding it quite palatable, with the flavor of hen's eggs. Pauraques' eggs are evidently not among those which have an unpleasant flavor that presumably protects them from predators. The latter are doubtless legion, including many ground-foraging mammals, snakes, and such terrestrial birds as domestic chickens. With so many hazards to reproduction one wonders how the Pauraque manages to remain abundant over so wide a range. Probably the adults live a long while and can make many nesting attempts in the course of their lives.

Summary

One of the most widespread and common birds of tropical America, the Pauraque inhabits both humid and semiarid regions from sea level up to 5,000 or 6,000 feet and occasionally even higher. Avoiding heavy forest, it finds optimum conditions where woodland edge, thickets, canefields, or the like alternate with open spaces. By day it rests on the ground beneath sheltering vegetation; at nightfall it comes out into the open to catch flying insects.

The Pauraque calls, or sings, chiefly in the dry season. With a full moon it sings through much of the night; when the night is moonless it sings in the evening and morning twilight.

The Pauraque breeds in the drier part of the year. North of about 5° north latitude the nesting season extends from February or March until May, June, or, rarely, July.

The set regularly consists of two eggs which are laid directly on the ground litter or the ground itself, with no attempt to build a nest. Often they are placed in a spot somewhat more open than that in which the Pauraques rest by day. In two instances the second egg of a set was laid late in the afternoon, long before the Pauraques became active in the twilight. The eggs, light buff or pinkish buff spotted and blotched with shades of brown and lilac, are usually considerably lighter than the substratum and much more conspicuous than the Pauraques themselves. But because, from the time the first is laid, they are almost constantly covered by the parents, they seem not to require better cryptic coloration.

Both sexes incubate. In the daytime the male covers the eggs more than the female and sometimes he sits throughout the day. Considerable individual, and perhaps also geographical, variation in the amount of diurnal incubation performed by the female has been noticed. For short intervals in the evening and morning twilight the eggs are left exposed while both parents forage. Nocturnal incubation seems to be performed chiefly by the female, but observation is difficult in darkness. While Pauraques incubate in daytime their eyelids are in constant movement, between closed and half open. Rarely are the eyes fully open and rarely do they remain closed for more than a few seconds.

Pauraques have not been known to transport their eggs in consequence of human interference. But when, to save them from fire ants, I shifted the eggs a short distance, the birds moved them back, probably by rolling, to their original position over the ants' nest with the result that the ants killed the chicks.

Newly hatched Pauraques are almost completely covered with soft down and their eyes open very soon after hatching. From their first day they can move by hopping over the ground, which they do in response to special notes of the parents, who settle on the ground some distance away and call the chicks to themselves. Even when only a day or two old they have a surprising capacity to surmount obstacles.

The chicks are fed by regurgitation, the parent inserting its small bill into their mouths. Because of the mobility of the chicks from the day they hatch, it is most difficult to learn much about feeding.

Parents continue to attend young until they are well feathered and can fly.

Parent Pauraques often give distraction displays, grovelling on the ground and beating their wings against it. Such displays are given before the eggs hatch but are more frequent afterward; they continue until the young can fly well. One pair lured an opossum from their eggs by means of repeated distraction displays. Individual Pauraques differ greatly in their readiness to give these displays.

Among the causes of loss of eggs and young are fire ants, ground fires, trampling by large animals, and doubtless predation by a great variety of reptiles, mammals, and birds.

Family TROCHILIDAE

BAND-TAILED BARBTHROAT

Threnetes ruckeri

In its long, strongly curved bill and lack of the glittering plumage typical of hummingbirds, the Band-tailed Barbthroat resembles the hermits *(Phaethornis* spp.) which, like itself, inhabit the understory of tropical woodland. Sometimes it is known as "Rucker's Hermit." It is about 4½ inches long. In both sexes the upper parts, including the middle pair of feathers of the rounded tail, are metallic bronze-green. The outer tail feathers are white at the base, dull black in the middle, and white or pale buff at the end; when spread, they display three broad, contrasting bands. A dusky patch covering the cheeks and ear coverts is margined above and below by buffy stripes. The chin and upper throat are dusky and the chest is cinnamon-rufous, which fades to buffy gray on the abdomen. The upper mandible is black and the lower pale yellow. The eyes are dark, the legs and toes pale flesh-color.

The barbthroat inhabits tall, lush thickets and the undergrowth of humid forests from Guatemala to Colombia and Venezuela. In Costa Rica it ranges altitudinally from sea level to at least 2,500 feet. Like other hermits it is almost fearless of man. Sometimes, while I have made my way toilsomely through tangled undergrowth, one of these long-billed birds has flown up to hover all around me, bill toward me, within arm's length. Fanning my heated cheeks with its wings, it has scrutinized me in a manner that made me suspect that, like other hummingbirds, it is nearsighted—which would not be surprising considering the close range at which these birds pluck tiny creatures from foliage or snatch volitant insects from the air. Nor is the barbthroat averse to being scrutinized in turn; occasionally, while perching one has permitted me to examine it with my eyes hardly a foot from its own.

A substantial portion of the barbthroat's food consists of small insects, spiders, and the like, gathered from beneath the leaves of bushes and low trees in the undergrowth, while it hovers below them. It is often found among stands of such great-leafed herbs as wild plantains *(Heliconia* spp.), shellflowers (*Calathea* spp.), and related genera. At the end of December 1930 I studied the floral behavior of a shellflower *(C. lutea)* on Barro Colorado Island, in Gatún Lake in the Panama Canal Zone. In the sunshine on the lake shore, at the edge of a small banana plantation and not far from the forest, grew a luxuriant cluster of these herbs with broad, upright leaves on long stalks. Day after day a barbthroat spent much time on a slender dead vine among the shellflower leaves. As it rested there, usually facing me, its white-tipped tail was constantly wagging up and down, like the tails of singing Little

Hermits and Long-tailed Hermits, although this bird was silent. Again and again it protruded its long, forked, white tongue. Although it frequently visited the inflorescences of the shellflower, it did not draw nectar from the small yellow flowers, the tubes of which were too narrow to admit its bill. It probed the water that had collected inside the large, fleshy, two-ranked, folded bracts that embraced the flowers, but whether to sip the liquid or to devour insect larvae which flourished there I could not learn.

Voice and Courtship

Like many other hummingbirds, barbthroats, presumably males, announce their presence and availability to the other sex by persistent song, delivered day after day in a definite spot. The stations of those that I have watched performing so were in the tangled undergrowth of the rainforest. Here, where light was dim and visibility narrowly restricted, the singing barbthroat rested on a slender twig, often a dead one, a yard or two above the ground. The first that I found, in July 1941, was alone and in the next two years he continued to be the only songster of his kind audible in the locality. But in April 1944 a second barbthroat was proclaiming his presence not far from the first, who for the fourth consecutive year was occupying the same station; and by 1947 at least three were singing in this part of the forest. The two farthest apart were about 250 feet from each other. Although some hermits, particularly the Little Hermit, form more populous courtship assemblies, this group of three was the largest gathering of barbthroats that I have found. Perhaps at lower altitudes, where barbthroats are more numerous, the males join in larger groups to compete for the females. In this same small area of forest where I first heard a barbthroat in 1941, they continued to sing until 1956, or for 16 years. A similar perpetuation of courtship assemblies has been noticed in several other species of tropical hummingbirds.

Although only technically, or by courtesy, can many hummingbirds be said to sing, the Band-tailed Barbthroat is a true songster. His voice is thin but his verse is surprisingly varied and tuneful, one of the few hummingbird's songs that seems to be delivered with feeling rather than mechanically. Indeed, when I first heard a barbthroat I suspected that I listened to a warbler or some other small songbird, and I was not convinced of my error until I had traced the notes to their source. The song is plaintive, suggestive of melancholy; and often the notes follow each other so rapidly that they almost form a trill. Each song lasts about four or five seconds and then one must wait a much longer interval for the next repetition. The barbthroat does not squander his songs like hummingbirds less musically gifted, but he spaces them out through much of the day. While on his song perch he incessantly waves his tail up and down through a longer or shorter arc.

When occasional showers keep the forest damp through the early months of the year, barbthroats may begin their season of song in late March; if the dry season is long and severe, singing may not start until early May. These hummingbirds continue to perform at their usual stations through the early half of the rainy season, until mid-August or a little later, then fall silent until the following year. This applies to the elevated Valley of El General; in the Pacific lowlands near the Golfo Dulce I heard two or three barbthroats singing persistently, within hearing of each other, in the dark undergrowth of the forest on 24 December 1947.

Barbthroats have continued to sing while I stood a yard away. One day I noticed that the songster's long bill remained almost closed while his mouth opened and closed. The movements of the lower mandible did not follow or correspond to those of the lower jaw, a situation that could arise only if there were a joint or point of flexure between the two. Brauner (1953:72) described a "hinge" in the short lower mandible of the Poor-will, which permits this goatsucker to call without appearing to open its huge mouth, thereby imparting a ventriloquial quality to its notes. Evidently the barbthroat's much longer lower mandible is similarly hinged, but relatively nearer the base.

Nest and Eggs

So many bunches of bananas and plantains were stolen from our plantation beside the boundary of the farm that, when the time came to make a new planting, we hid it in the midst of tall secondary forest on low rocky ground near the river. This little plantation surrounded by woodland and luxuriant stands of wild plantains proved to be especially attractive to hummingbirds. Here, 17 months after the bananas were set in the ground, I found the first nests of the Band-tailed Barbthroat and the Bronzy Hermit that I had ever seen. Each of these hummingbirds nested here again in the following four years. While watching them I saw 10 other species of hummingbirds visiting the banana flowers or flying through the plantation.

The seven barbthroat's nests built in this plantation from 1967 to 1971, possibly all by the same hummingbird, were quite similar in site and construction. Each was attached to the under or inner side of a strip of a living banana leaf torn from margin to midrib and hanging almost vertically below the latter, at heights of 10 to 13 feet above the ground. The green ribbon of leaf tissue to which each nest was attached was about 1½ inches wide. The nest, which stood out from the leaf strip like a concave bracket, was a rough, stiff, springy fabric, made largely of thin, wiry, richly branched stems of liverworts and mosses from which the leaves had largely fallen, or else had dried and shrivelled. There were a few black fungal rhizomorphs, or "vegetable horsehairs," especially coiled down in the bottom of the cup, which could hardly be said to be lined. A few small pieces of light gray foliaceous lichens were attached to the outer surface, and there were some fragments of green moss. Although the walls were rather thick, their loose construction permitted much light to pass through. The nest was fastened to the strip of leaf chiefly by strands of whitish cobweb, which were especially prominent on the back of the leaf, but a few fungal filaments and liverwort stems were also wrapped around the strip. The loose "tail" hanging below the end of the leaf strip was composed of liverwort stems and shrivelled small leaves of seed plants. The prevailing color of the whole structure was deep brown to almost black. The total length of one was nine inches, of which 5½ were accounted for by the tail. The cup was two inches in over-all diameter at the top. The inside diameter was 1¼ inches, depth one inch. Another nest was similar in construction but only 3½ inches long, for it had scarcely any tail.

The first nest in the new banana plantation was nearly finished when found but I saw a little of the process of building. Arriving with material the barbthroat settled in her nest with her breast toward the supporting strip but turning frequently from side to side as she shaped the structure. While working she continued to beat her wings rapidly. She was wary and refused to approach her nest if I stood closer than about 60 feet. This shyness was surprising in a bird that sometimes hovers at arm's length from a man's face as he walks through the woodland.

On 6 August 1967 the first egg was laid in this nest. The second egg was deposited two days later. After a week both eggs mysteriously vanished. On 18 August I discovered a new nest at the opposite corner of the banana plantation, about 50 yards from the first. It already contained one egg and the second was laid two days later. If, as was probable, this nest belonged to the barbthroat who lost the preceding nest, it was built, and the first egg laid in it, in less than a week. This nest was successful and the following observations on incubation and care of the young were made at it.

On 11 June 1968 there was another nest in the banana plantation, very near the site of the first. It already contained one egg and another lay broken on the ground below. While one barbthroat sat in the nest another darted up and hovered beside it, then quickly flew away. This was repeated about a dozen times. After the incubating bird left, the other barbthroat, as I infer, came repeatedly to hover in front of the nest and apparently look into it. The one who had evidently built the nest brought two stiff fibers and some cobweb which she wrapped across the leaf strip while circling on wing around it.

By 14 June this nest held three eggs. Including the fallen egg, four had been laid, undoubtedly by two females, both of whom were bringing material to it. Thrice, while one incubated, the other flew up with cobwebs or fibers in her bill and tried to add them to the nest, only to be held aloof by the open, threatening bill of the sitting hummingbird. While the eggs were unattended material was placed in it four times

FIG. 6. Nest of Band-tailed Barbthroat, attached to strip of banana leaf, with three eggs laid by two females. Los Cusingos, Valley of El General, Costa Rica, June, 1968.

left. A minute later a barbthroat, apparently the same individual, again settled on the eggs only to leave in the following minute, pursued by the other.

At this nest incubation continued until 4 July, or for at least 20 days, after which the eggs were abandoned. Five days later I measured, then opened them. Two had no trace of an embryo; the third, a very small one, about 2 mm. long. Surprisingly, no trace of the odor of putrefaction came from any of them.

These three eggs measured 15.0 by 9.0, 15.2 by 9.0, and 15.3 by 9.3 mm. Although scattered through the literature there are a few records of hummingbirds' nests, of various species, with three eggs, this barbthroat's nest, in which the three eggs were the product of two females, is the only hummingbird's nest with more than two eggs that I have ever seen. In a normal set of two barbthroat's eggs, they measured 14.5 by 9 and 14.3 by 9 mm.

From the first nest built in 1970, the last young left on 12 August, slightly prematurely, for the structure had turned sideways. About 100 feet from this nest I found, on 3 September, another nest already with two eggs, which apparently represented a second brood although I was not certain that the young of the first brood survived. These eggs were incubated for at least 17 days, then abandoned after an afternoon of torrential rain on 20 September. Opened a few days later, they contained dead embryos nearly ready to hatch. In seven nests in this plantation eggs were laid from mid-June until the beginning of September, and the latest young flew in early October. Outside this interval I have searched in vain for barbthroats' nests among the banana plants, where they are not hard to find when present. Like the Bronzy Hermit nesting in this plantation, the barbthroat breeds only in the rainy season.

Incubation

My son, Edwin, and I watched the second nest from daybreak until noon on 29 August 1967 and throughout the following day. Evidently only the female incubated the two eggs, for we never saw more than one barbthroat near this nest. On 30 August the female returned to her eggs at 14:07 and continued to sit until nightfall, during the late afternoon under intermittent showers. This early beginning of the long nocturnal session greatly curtailed the period during which diurnal sessions with intervening recesses were observed. Taking both days together, we timed 11 such sessions, ranging from three to 180 minutes in length and averaging 64 minutes. Late in each forenoon the barbthroat took a very long session, continuing from 09:01 to 12:01 on 29 August and from 09:49 to 11:35 (106 minutes) on 30 August. Early in the afternoon of this day she sat continuously for 97 minutes. Only two sessions, in the early morning, were shorter than half an hour. Eleven recesses ranged from four to 30 minutes in length and averaged 17.4 minutes. Calculated in the usual manner, this barbthroat incubated with a constancy of 79 per cent during her active period. But if we consider the total elapsed time on 30 August between 05:30 and 17:45, when I left in the rainy dusk, she covered her eggs for 83 per cent of the day.

The barbthroat always sat with her breast inward toward the strip of leaf to which her nest was attached, her head thrown far back and her bill held almost straight upward. Her posture while incubating was exactly the same as that of other hermits *(Phaethornis, Glaucis)* and gave the same impression of an almost unbearable flexure of the body; but doubtless the hummingbirds do not find uncomfortable this attitude which they are able to maintain for hours together. Mostly the barbthroat sat motionless, but from time to time she turned her head around to preen the ends of her wings, her rump, or her tail. At intervals she opened and closed her bill mincingly. Sometimes these bill movements would be followed by tail-wagging. Then the barbthroat would spread her wings and fly backward until she hovered with the nest in front of her, the tip of her bill in the cup. She did not appear to turn her eggs but to inspect them. After a moment in this position she darted away.

This barbthroat brought material for her nest a few times in the early morning. Sometimes it appeared to consist of fine rootlets or similar vegetable matter, rolled up into a loose ball, which after settling in the nest she placed loosely on the rim beside her or against the leaf in front. She did not once spread cobweb over the back of the supporting leaf to reinforce her nest's attachment to it, as hermit hummingbirds so frequently do. But at the third barbthroat's nest this precaution was not omitted. Once, while spreading cobweb over the leaf and the outside of this third nest, the hummingbird made seven revolutions around it, hovering on wing and keeping the point of her bill close to the leaf and nest. These revolutions were made in both directions; after making some clockwise turns she reversed to revolve counterclockwise.

In the second nest the first egg hatched on 7 September, between 07:30 and 12:45. Arriving at the latter hour I watched the parent return and, hovering in front of the nest, feed the newly hatched nestling, as I could see through the meshes of the wall. Then she took the larger part of the divided shell in the tip of her long bill and started to fly off with it. But after going a foot or so she dropped it, caught it in the air, carried it a foot or two more, all on a descending course, then dropped the piece of shell to the ground. Promptly she returned to pick up the cap of the shell and bear it a short distance away. Then she came again to brood.

The second egg in this nest hatched between the evening of 7 September and 07:10 next morning, 18½ to 19 days after it was laid. At another nest, the incubation period was at least 18 days. This is an extraordinarily long incubation period for a hummingbird. For those that build cup-shaped rather than hanging nests the period is usually 15 to 17 days, although it may be prolonged a few days by unfavorable weather. The barbthroat sat with high constancy; the weather while she incubated was not particularly inclement for the wet season; and her nest was protected from the heavy rains by the banana leaf beneath which it hung. Probably the open fabric of the nest, which afforded poor insulation for the eggs, contributed to the length of the incubation period. The situation of the nest was probably also not without influence: in other families, such as the American flycatchers (Tyrannidae), species with pensile nests have substantially longer incubation periods than do those with more conventional nests (Skutch, 1960a:575). But in the Little Hermit, whose nest fastened beneath a leaf is well padded with down, the incubation period is only 15 or 16 days.

Nestlings

Soon after they hatched, the nestlings oriented themselves with their heads inward toward the supporting leaf, just as the parent incubates, and they preserved this orientation until they flew.

From 05:30 to 11:50 on the morning of 10 September, when the two nestlings were two days old, their mother came eight times to feed them. Clinging to the nest's outer rim with raised, rapidly beating wings, she seemed to regurgitate to them alternately, but so little could be seen of the tiny nestlings that it was difficult to discern details. Each meal was followed by brooding and once the parent came to brood without feeding. Her nine sessions of brooding ranged from nine to 45 minutes and totaled 234 minutes. An equal number of absences ranged from 12 to 23 minutes and totaled 146 minutes. The parent brooded for 61.5 per cent of the morning, always sitting with her head inward as she did while incubating.

Before proceeding to the nest to feed or brood the parent always rested, a yard or so above the ground, at the edge of the thicket nearest the nest. While perching here she constantly wagged her tail rapidly up and down and usually she preened vigorously, lifting her pale flesh-colored leg over a drooped wing to scratch her head or bill. Then she flew up to the nest beneath a banana leaf.

On 19 September, when the nestlings were 11 days old, we watched from 05:32 to 11:32 on a sunny morning. The parent came with food nine times. Clinging to the lowest part of the rim with rapidly beating wings, as before, she reached over the nestlings' backs to insert her bill into their uplifted mouths, which were now visible above the side of the nest. On every visit each nestling was fed in two or three installments or separate acts of regurgitation, between which the bill was removed from the recipient's mouth. Usually they were fed alternately. They were not brooded this morning.

On 28 September, when the nestlings were 20 days old and well feathered, we watched from 05:25 to 11:30, again on a clear morning. The parent brought food 12 times, usually delivering it in three installments, two to one nestling and one to the other. Now they lifted their heads very high and sometimes their mother fed them while hovering in the air, as though she were sucking nectar from a flower. Once she fed one nestling while clinging to the head of the other, her wings rapidly beating. Then she clung to the rim at the side of the nest to feed the nestling whose head she had used for support. The young hummingbirds sometimes stood on the rim while they flapped their wings into a haze, holding on to prevent being borne away.

On 29 September the parent brought food at the most rapid rate we had so far recorded, five times during the first hour of the day; but sometimes each nestling received only a single installment.

During the second half of the nestling period a Long-tailed Hermit often perched at the thicket's edge, just where the barbthroat rested before going to the nest. Frequently the hermit hovered in front of the nest, looking at the nestlings and sometimes touching them lightly with the tip of its long bill. Once it seemed to feed the nestlings. Less frequently a Bronzy Hermit appeared, and once I saw it hover in front of the nest much as the Long-tailed Hermit did, without attempting to feed. The parent barbthroat was not seen to chase these other hummingbirds. Hermit hummingbirds, including barbthroats, display great interest in each other's conspicuously hanging nests and doubtless this explains why two females sometimes lay in the same nest.

At midday on 1 October both nestlings were still present. Twenty-four hours later the cup or bracket of the nest had either fallen away or been torn away by some animal, more probably the latter, since I could find no piece of it on the ground below. The culprit was doubtless some flying thing, as no animal large enough to tear the nest could have climbed to it without leaving revealing signs on the banana leaf. Possibly it was a bat. The nestlings had vanished. One was 24 or 25 days old, the other 23 or 24. They were sufficiently developed to fly, if they escaped alive.

Young after Leaving the Nest

I was not sure what happened to the nestlings until three weeks after their disappearance. Then, on 23 October, I found a juvenile in a thicket about 50 yards from the site of the fallen nest. Considering the rarity of barbthroats in this region and the age of the young bird, I had no doubt that it had been raised in the nest that we had watched. About full grown, it had a tail much as in the adults but its under parts were pale gray, with no cinnamon on the chest. It rested on slender horizontal twigs, three or four yards up, from time to time shifting from one to another but always returning to the same small area after each brief excursion. Here its mother came and fed it twice, regurgitating to it while the two perched side by side.

The juvenile hovered before some small, pale red leaves at the end of a young vine shoot as though it had mistaken them for flowers and tried to draw nectar from them. It also hovered in front of, and tried to stick its bill into, the expanded blackish base of the petiole of a dead cecropia leaf caught up in the thicket. It did not appear to pluck insects from these nectarless objects. Evidently it had not yet learned to distinguish flowers from other things that in shape or color somewhat resembled them. It seemed still to depend largely on its parent for food.

Since young hummingbirds do not, like many other kinds of birds, follow their foraging parents, it is important for them to have a definite spot where they are fed, to which both return after their independent excursions. Without such a rendezvous they might lose contact with each other, to the great detriment of the dependent juvenile. I first became aware of this facet of hummingbird behavior in the Scaly-breasted Hummingbird (Skutch, 1964b:195-196).

Morning after morning for the next 10 days I watched the young barbthroat being fed in the same place. Usually the parent arrived with food from two to four times in the hour from about 07:00 to 08:00. After receiving a liberal meal while the two perched side by side, the juvenile often followed its mother when she started to fly away, causing her to alight on another perch a few yards from the first and regurgitate to it again. Sometimes the pursuit was repeated and the young bird was fed once more on a third perch. At times, after receiving the last installment, the young bird flew off out of sight with its parent. But nearly always it returned before she did and waited for her in the thicket. While waiting it tugged at small decaying leaves or bits of dead stem, but it seemed to find nothing edible in them. It also billed projections on stems. As days passed it spent more time hovering beneath green foliage, seeming to glean insects or spiders too small for me to see.

This young barbthroat was so fearless of me that I could come within a few inches of touching it while it perched at the edge of the thicket just above my reach. While resting on a slender twig it wagged its tail vigorously up and down. It stretched both half-opened wings upward above its back and also downward along its sides, with the primaries partly spread. At the same time the tail was fanned out conspicuously, revealing the white bases of the lateral rectrices. It scratched its head by raising a foot above a relaxed wing, as adult barbthroats do. Sometimes it exercised its wings, beating them vigorously while clinging to its perch. When hungry it called with single sharp notes. One morning an adult barbthroat hovered in front of the juvenile, displaying with its tail spread to show the bold, black-and-white-banded pattern. When the juvenile flew to another perch the adult followed, to display before it once more. Then the two rested a while close together and hardly a yard from me.

On 26 October, when the juvenile was seven weeks old, I noticed that a small tawny patch had appeared in the center of its gray throat. The last time that I saw it receive food was on 3 November, when it was 56 days old and had been out of the nest for 32 days. A young Scaly-breasted Hummingbird was fed until it was at least 65 days old and had been on the wing for 41 days. Another Scaly-breast was attended by its parent for at least four weeks after it left the nest, or until it was 52 days old. A young Long-billed Starthroat was fed for no less than 23 days after its first flight, or until it was 48 days old. A White-eared Hummingbird received food from its mother, who was incubating a second set of eggs, for at least two weeks after it took flight, or until it was 40 days old. More recently, I watched a Rufous-tailed Hummingbird that was fed for more that a month after it flew from the nest, until it was 59 days old.

Summary

The Band-tailed Barbthroat inhabits tall, lush thickets and the undergrowth of humid forests. Small insects and spiders gleaned from the undersides of leaves contribute largely to its diet.

Perching low amid the forest undergrowth, males repeat a plaintive, long-continued song. Sometimes several form a small courtship assembly. In the Valley of El General singing may begin in March in wet years but not until late April or early May when the dry season has been long and severe; it continues into August. One song station was occupied for 16 years.

Occupied nests were found only in the rainy season, from mid-June until early October. Each was fastened with cobweb to the underside of a narrow ribbon of green tissue formed by the splitting of a large, living banana leaf.

Each of six nests contained two eggs but one nest held three eggs, with another on the ground below. These four eggs were evidently laid by two females. While one incubated the other often hovered beside the nest, and sometimes she brought material to it.

The incubating female always sat facing the supporting leaf, with her head thrown so far back that she was bent almost double, as is typical of hermit hummingbirds. Her sessions were rarely shorter than half an hour and sometimes lasted three hours; she incubated with a constancy of about 80 per cent. The incubation period was 18½ to 19 days. The parent carried away the empty shells.

Soon after hatching, the nestlings oriented themselves with their heads toward the supporting leaf, which position they maintained until they flew. In the first six hours of the morning their mother brought food to the two nestlings eight times when they were two days old, nine times when they were 11 days old, and 12 times when they were 20 days old. On nearly every feeding visit both nestlings were fed once or twice, alternately.

The young remained in the nest for 24 or 25 days. One juvenile continued to be fed for at least 32 days after leaving the nest, or until it was 56 days old. All observed feedings occurred within a radius of a few yards in a thicket. Since fledgling hummingbirds do not follow their foraging parent it is important that they have a definite place where mother and dependent offspring meet.

BRONZY HERMIT

Glaucis aenea

The Bronzy Hermit is a rather dull-colored, medium-sized hummingbird about four inches long. The top of its head is dusky and the rest of its upper plumage is metallic bronze, with cinnamon margins on the upper tail coverts and dusky wings. There is a broad, black band behind each eye. The under parts are rusty cinnamon, becoming paler on the chin. The central feathers of the rounded tail are bronze-green; the others are rufous-chestnut; and all have white tips and black subterminal bars that are broadest on the outer feathers. The long, strongly curved bill has a black upper mandible and a dull yellow, dusky-tipped lower mandible. The eye is dark and the legs and toes are pale red, pale orange, or flesh-color. The sexes are nearly alike but on the female the cinnamon of the under parts is somewhat brighter and more uniform.

This hermit has a discontinuous range. Typical *aenea* is found from Nicaragua to western Panama; the species is unknown in eastern Panama; and the race *columbiana* extends from western Colombia into northwestern Ecuador (Wetmore, 1968:255-257). Although known chiefly from the lowlands, the Bronzy Hermit has been found nesting as high as 4,000 feet in the Province of Chiriquí on the Pacific slope of western Panama (Worth, 1939:307). It seems chiefly to inhabit tall, lush second-growth vegetation, from which it enters adjoining shady plantations. I never saw this hummingbird until I found it nesting at the edge of the same small banana plantation where the Band-tailed Barbthroat nested.

Nest and Eggs

The large leaves of wild plantains (*Heliconia* spp.) tear into transverse strips much in the manner of the related banana although, being mostly smaller, this does not occur so frequently. The long, slender leaves of a Heliconia with great, dangling, red inflorescences leaned over the edge of the little banana grove close beside a rivulet. Near the end of one of these leaves, attached beneath an inch-wide strip of leaf-blade, my first nest of the Bronzy Hermit hung 20 feet above the ground. The structure consisted of a moderately deep cup, fastened by cobweb to the ribbon of leaf that formed a roof above it, and a long, slender "tail" that dangled free below the end of the strip. The body of the nest was a stiff, springy fabric so loosely constructed that much light passed through its walls. These were composed of richly branched stems of liverworts from which the tiny leaves had fallen or on which they were dry and brown, black fungal rhizomorphs, fine fibrous rootlets, and similar materials, all of a dark color. The fibers in the bottom were somewhat finer but there was no distinct lining. A few greenish gray lichens were fastened to the outside. The dangling tail contained some fairly large, shrivelled dead leaves. The cobweb that held the nest in place formed a conspicuous coating over the upper surface of the strip of leaf. The body of the nest was three inches high, 1¾ inches from side to side, and two inches front to back. The inside measurements were one by 1½ inches in diameter by 1¼ inches deep. The nests of the Bronzy Hermit were more slightly built, with thinner walls, than those of the Band-tailed Barbthroat in the same plantation.

This nest was found on 2 August 1967 when it held newly hatched nestlings. In each of the four subsequent years a single Bronzy Hermit has built in this same area. Of the five nests found here, the first two were attached to wild plantain leaves at the edge of the plantation and the next three to narrow strips of banana leaves in the plantation. The height of these nests ranged from 12 to 20 feet. The most distant of the nest sites were about 50 yards apart.

The second of these nests, built in 1968, was double. I did not find it until it already contained two young, which flew about 3 July, much earlier than any other brood that I have seen. On 15 July I noticed that a new nest was being built on the same narrow strip of Heliconia leaf, immediately above the first and close to the midrib to which the strip was attached. By 4 August this higher nest was finished and contained two eggs. Three nests of the Bronzy Hermit found by Worth (1942) in a plantation in Chiriquí Province, western Panama, were attached beneath inch-wide strips of banana leaves about 10 feet above the ground.

As in other hummingbirds, the nest is built by the female alone. The attachment of the earliest pieces to the slippery strip of leaf, which I have not watched in this species, can only be done while she hovers on wing. At a later stage, when the bowl is taking shape, she alternately sits in it and floats around the back of the leaf strip, facing it, while she applies more cobweb. The thin, wiry pieces that form the bowl are sometimes recalcitrant. With her long bill she repeatedly tucks them down into the growing nest, and often she uses her feet to knead and shape the hollow in which she sits, as can be seen through the open fabric. Sometimes hovering on wing she plucks, or tries to pluck, fibers projecting from the bottom of the nest to place them in the cup. In the bright sunshine from 07:00 to 08:00 on 6 July a female brought material 14 times. On the following morning, which was cloudy, she did so only seven times in the same interval. In the following year this female brought material to her nest 13 times in the hour from 07:10 to 08:10. In both years she was attended by a male, as will be told beyond.

The construction of the nest is evidently a lengthy process. I have not found a nest in the earliest stages. In two years a period of six days elapsed between my discovery of the nest, which was already far advanced, and the laying of the first egg. At one nest the second egg was laid before 07:00, two days after the first egg. At another nest an interval of four or five days separated the laying of the eggs. Possibly an egg had been lost and the female replaced it to complete her set of two.

In five years six sets of eggs in and beside this small banana plantation were laid from late May to early August, and the latest nestlings, had they survived, would have flown in mid-September. Outside this interval I searched fruitlessly for these conspicuously placed nests. Worth found eggs in Panama in June and July. Thus the breeding season of this hummingbird coincides closely with that of its neighbor of similar habits, the Band-tailed Barbthroat. Both breed in the wet season. It is significant that their eggs and nestlings are well protected from the frequent downpours by the strip of leaf to which the nest is fastened.

Worth's three nests, like the six that I found, each contained two eggs or nestlings. He gave the measurements of one set as 15 by 9 and 16 by 9 mm.

Only in 1968, when an unusually early, successful first nesting was followed by a second laying in the two-story nest, did I find evidence of a second brood. In this late nest only one egg hatched, and the nestling vanished when a few days old.

Incubation

At the second brood in the two-story nest beneath the Heliconia leaf in 1968, I watched the parent incubate throughout the morning of 17 August. She took 12 sessions, ranging from six to 42 minutes and averaging 19.5 minutes. An equal number of recesses ranged from four to 21 minutes and averaged 12.1 minutes. The hummingbird covered her eggs only 61.7 per cent of the morning. She incubated much less constantly than her neighbor the barbthroat, who even in the forenoon rarely sat for less than an hour at a stretch and once incubated continuously for three hours. Possibly this Bronzy Hermit was still feeding the young of her previous brood, who had flown from the lower bowl of the two-story nest six or seven weeks earlier, although this would have been unusually prolonged parental care. She failed to bring material to the nest, as incubating hummingbirds commonly do in the mornings.

The hermit always sat on her eggs with her breast toward the supporting strip of leaf, body doubled up, head thrown far back, and bill held straight upward, in the seemingly strained posture typical of hummingbirds that attach their nests beneath leaves. Like them she left her nest by flying upward and backward, then reversed to dart away.

On 30 June 1971 I again passed a morning watching a Bronzy Hermit incubate at a nest only a few yards from the site of the two-story structure. From 05:30 to 11:32 this hummingbird took 16 sessions, ranging from four to 24 minutes and averaging 13.1 minutes. Her 16 recesses varied from four to 20 minutes in length and averaged 9.5 minutes. She sat on her eggs only 58 per cent of the six hours and she brought no material to her nest. Perhaps she sat so inconstantly because she was compensating for meager fare on the preceding afternoon, when almost continuous hard rain had doubtless curtailed her foraging.

At this nest the incubation period was between 17½ and 19 days. At one earlier nest it was at least 17 days and at another it was between 17 and 17½ days. In the closely related, possibly conspecific, Rufous-breasted Hermit, Novaes and Carvalho (1957) found the incubation period to be 16 days at Belém, Pará, Brazil. Probably the higher air temperatures at sea level in this equatorial locality account for an incubation period shorter than that of the Bronzy Hermit and the Band-tailed Barbthroat in the elevated Valley of El General.

Nestlings

The second nestling hatched one or even two days after the first. On several occasions I have found the empty shells lying on the ground beneath the nest, where they had evidently been dropped by the parent, instead of being left in the nest to be crushed beneath the nestlings and their brooding mother, as happens in some hummingbirds.

Brooding.—During the first six hours of the morning, nestlings two and three days old were brooded for 15 intervals ranging from five to to 15 minutes and averaging 9.7 minutes. Fifteen absences varied from one to more than 55 minutes and averaged more than 13.4 minutes. Only absences of 14 minutes or more were followed by a feeding. These tiny nestlings were brooded for 42 per cent of the clear morning. During six hours of a rainy morning in the following year, nestlings two and four days old were brooded for 14 intervals, ranging from seven to 27 minutes and averaging 13 minutes. Their mother's 13 absences varied from two to 24 minutes and averaged 13.1 minutes. Accordingly, the nestlings were brooded for 50 per cent of the morning. Absences of 10 minutes or more were followed by a feeding, but after a shorter excursion the parent returned to brood without delivering food. Nestlings 14 and 15 days old were no longer brooded on a clear morning.

Feeding.—Soon after they hatch, the dark-skinned, blind, almost embryonic nestlings orient themselves with their heads toward the strip of leaf to which their nest is attached, in the manner of other hermit hummingbirds, and they maintain this position until they fly. To feed them their mother clings to the rim behind them, her flesh-colored legs stretched up conspicuously, her wings beating into a haze, her tail swinging up and down through a wide arc. Reaching over the nestlings' backs, she plunges her long bill into their throats and regurgitates to them, usually feeding each several times, alternately, on each visit.

As days passed the highest nest sank under the weight of its growing occupants until the opening, which at first had faced obliquely upward, came to face sideward; the rim was vertical. The nestlings' lengthening tails projected over the lower edge of the rim where the parent formerly clung to deliver food. Now she fed while hovering beside the nest, as she had sometimes done even before it sank down, or else she clung to the rump of one nestling while she regurgitated to the other. The young birds bent their heads far backward to receive her bill into their mouths.

During the first six hours of 30 July 1970 the female brought food seven times to nestlings two and three days old. On 10 July of the following year nestlings two and four days old were fed eight times in the first six hours of the day. When 14 and 15 days old the nestlings in the former nest were fed 10 times in the same interval. At the first nest, in 1967, my watches were shorter but more frequent. When the nestlings were only a few days old they were fed nine times in 5½ hours and in the middle of the nestling period they were fed eight times in 2¾ hours. Thus, the rate of bringing food to the nest increases from about 1.2 times per hour during the first few days to 1.7 or even 2.9 times per hour when the nestlings are becoming feathered. The rate per nestling/hour is half of these figures.

Sanitation.—While the nestlings were still too tiny to eject their excreta beyond the nest's rim, their mother sometimes carried away their droppings in the tip of her long bill.

Departure.—At the first nest the two nestlings, which seemed only a day or two old when I first saw them in a mirror raised on a long pole on 2 August, left between 24 and 26 August. The last to go was in the nest no less than 24 days. Novaes and Carvalho found the nestling period of the Rufous-breasted Hermit to be 22 days.

At a later nest I witnessed the departure of the last fledgling, whose exact age was unknown. After receiving a meal at 07:16 on a bright morning, the young hummingbird became very active, looking around much, beating its wings into a haze while clinging to its nest beneath a banana leaf. A few minutes later it rose from the nest and flew about 20 feet to the edge of the thicket beside the plantation. Already wary, it retreated deeper into the thicket when I approached. The parent was not in sight when the fledgling abandoned the nest.

The nesting in 1970 had an unhappy but instructive ending. One nestling disappeared when about 20 days old; whether it lived or died, I do not know. Its nest-mate lingered in the nest 10 days longer. This nest was attached to a strip only ⅝ inch wide, close to the tip of a great banana leaf where the massive midrib became narrow. When the two sides of the leaf folded downward beneath the midday sun, the nest was enclosed in a tent that made it difficult to see from the ground and for the parent to reach. Before the young left, the leaf, dying from the apex inward, became permanently deflexed around the nest. When the remaining nestling took the customary vigorous wing exercises its right wing struck against the leaf blade beside it. Soon the young hummingbird formed the habit of whirring its wings with only the left one fully extended; the right wing was kept partly closed and held somewhat back and up, to prevent its striking so hard against the dry leaf. Nevertheless, enough covert feathers rubbed off this wing to leave the skin exposed.

The mother continued faithfully to attend this nestling, sometimes feeding it three times in an hour, while clinging with rapidly beating wings to its back, wings, or lengthening tail. In addition to its frequent wing exercises, it preened much and scratched its head by raising a foot over its relaxed wing. Sometimes it rested in the nest with its head thrown far backward and bill pointing upward, like the incubating or brooding female.

When the nestling, 29 or 30 days old, still showed no inclination to fly, I tried to improve its situation by pushing out the part of the leaf blade which had bent so close to the nest that it interfered with the wing exercises. As the long pole that I used for this purpose neared the nest, the young hummingbird jumped out and fluttered to the ground at my feet. The underside of its right wing was raw and bleeding. Although the remiges of this wing were scarcely damaged, the young bird could flutter only a few inches. It could hardly perch but I finally managed to make it stay on a slender coffee branch with its breast resting on another twig.

Next morning the mother came to the place where I had left the nestling. Not finding it there she flew up to the empty nest, then back into the thicket. Without much doubt the fledgling had succumbed.

This episode demonstrated the importance of having a clear space around a hummingbird's nest so that the nestlings can exercise their wings without striking anything. Hermits do better when they attach their nest nearer the center of a leaf of the banana or wild plantain, for the apex is the first part to die, and as it dries it folds around a nest hanging beneath it.

Intruding Females

In most hummingbirds that have been carefully studied the breeding female leads a solitary existence, without intimate relations with others of her kind. With the Bronzy Hermit the case is different: she may be persistently molested by another female; she may be attended and guarded by a male; and both of these situations may arise at the same nest, leading to complex interactions difficult to interpret because the sexes cannot be distinguished while the birds dart rapidly about.

The whole time that the young of my first nest were growing up in their high nest beneath a Heliconia leaf, their mother was actively persecuted by another Bronzy Hermit, apparently another female, as I noticed no difference between the parent and her troublesome neighbor. The latter, whom I shall call the unemployed bird, spent much time perching on a slender, head-high twig just within the neighboring thicket, wagging her tail up and down in the manner of hermit hummingbirds. From time to time she flew up to visit the nest. If the parent were absent the unemployed bird would look in while hovering beside it, sometimes sticking in her bill, then fly away. Such frequent visits of inspection to a nest containing young, without either feeding or brooding, are not typical of parents attending nestlings; I could never convince myself that the unemployed bird performed either of these parental offices, although once there was a question of whether it was she who fed, and once she may have brooded.

While the parent brooded the other's visits to the nest continued, sometimes being repeated as many as a dozen times during a single session. Although sometimes the unemployed bird merely hovered beside the nest for a few seconds, seeming to menace the brooder by thrusting the point of her long bill toward her, often there was a closer encounter. Frequently the unemployed bird clung to the side of the nest in which the parent was sitting. To defend herself from these attacks the brooding bird turned her bill toward the assailant, sometimes bending her head back until it was almost upside down. Occasionally the two hummingbirds seized each other's bills briefly, but they seemed never to injure each other. Much fine twittering accompanied these encounters. Usually the pugnacious visitor left while the parent continued to brood, but sometimes both flew off together, the pursuit continuing until they were beyond view in the thicket. At times the bird returning to the nest was pursued by the other.

Feeding was sometimes interrupted by these attacks. Thus, at 10:04 on 4 August one hummingbird, evidently the parent, alighted on the nest's rim as though to feed the nestlings; but before she could begin the other attacked her from the rear, poking at her back and apparently pulling at her feathers, causing her to desist from feeding and to crouch down in the nest. The assailant returned again and again, often clinging to the side of the nest. They struck their bills together and the one in the nest lunged vigorously at her persecutor clinging beside her. The nest swayed widely during these encounters. Finally the assailant withdrew, only to return after an interval and resume her attacks. More often the arrival of the second bird during the delivery of a meal led to a swift chase rather than to a fight at the nest. After the interloper had been driven away, one, evidently the parent, returned to resume feeding.

The parent did not always wait to be attacked but sometimes entered the neighboring thicket to seek her rival. If the latter happened to be present one chased the other swiftly through the adjoining plantation, to the accompaniment of high-pitched twitters.

Despite interference by the second *Glaucis* the nestlings appeared to develop normally. They remained in the nest about as long as the Band-tailed Barbthroats, birds of about the same size, were in their neighboring nest.

An Attentive Male

It will be recalled that in 1968 the female Bronzy Hermit, after raising an early brood, built a second story to her nest and in it laid another set of eggs. While this hermit was attending her first brood her nest was sometimes visited by a second Bronzy Hermit, who hovered in front of it but was not seen to feed the nestlings. Later, while the upper story was being slowly built, the second hermit frequently visited it, from time to time even going through the motions of building although I never saw it bring any material. Now, while incubation was in progress, this second hermit spent much time perching within the edge of the neighboring thicket, about six feet above the ground and 20 feet from the nest. From its pale abdomen, I took it to be a male. Although I could not record his movements accurately while keeping watch on the nest, he seemed to spend about as much time resting in this one spot as the parent spent on the eggs. From time to time he sang or twittered in a high-pitched voice. He scratched his head "indirectly," by raising a foot over his relaxed wing. At intervals he flew up to hover beside the nest. When the female left for a recess he accompanied or pursued her, and the same sometimes happened as she returned to her eggs. Then, leaving the nest, he flew to his usual station in the thicket. Unless their flying together to and from the nest is to be interpreted as hostile chasing rather than escorting, I noticed no antagonism between these hermits, as was true at the nest in the preceding year. I did not see the male clash with the female.

In the following year I did not find the nest in the midst of the banana plantation until the young were half grown. In a few hours of watching I saw no Bronzy Hermit except their mother near this nest. In 1970 the nest was attached to a banana leaf only a few yards from the site of the two-story nest of 1968. I found this nest while it was being built and continued to watch it until the retarded nestling fluttered to the ground, as already told. Throughout this long interval of nearly two months a male Bronzy Hermit spent much time perching in a small coffee bush almost beneath the nest, from time to time flying up to look into it. Sometimes while the female was building he chased her as she came or went. Once the pursuit was long–continued, around and around through the banana plants, with shrill twitters. Finally one bird, evidently the female, sat in the nest and the other attacked her. Then, for a minute or two, he clung to the far side of the nest where I could not see just what he did. After this he returned to perch in the coffee shrub, whence from time to time, while the female was absent, he flew up to hover in front of the nest and poke his bill inside. Possibly when he appeared to attack the female on the nest that she was building he was merely trying to look inside, but was prevented from doing so by her presence. I did not see him bring anything to the nest.

During the period of incubation and raising the young the male was quieter, resting for long intervals below the nest, inspecting it less frequently than at first, and sometimes escorting, or pursuing, the female as she flew off to forage. I did not see him incubate or feed the nestlings; his role seemed to be to guard the nest. Sometimes he chased away hummingbirds of other kinds who came too close while seeking nectar from the banana flowers. Chiefly, however, he protected the female parent and her nest from a second female who from time to time flew out of the neighboring woods, exhibiting some of the aggressive behavior that I had witnessed at the first nest in 1967. But before she could seriously interfere with the incubating or brooding female, the male drove her away.

In 1971, when the nest was close to the site of that of the preceding year, the male rested frequently in the same coffee shrub, almost beneath this latest nest, which from time to time he flew up to inspect. Twice during building he actually sat in it momentarily. He also circled around the supporting strip of leaf as the building female does, although, as far as I saw, he made no contribution to the nest. Although he spent less time below the nest than in the preceding year, he continued to rest there at intervals until the eggs hatched. During six hours of the morning of 10 July, when the nestlings were a few days old, he came to the coffee bush 10 times and perched there for a total of 51 minutes. After this I was obliged to suspend observations.

Discussion

It can hardly be doubted that the nests in closely adjacent sites in 1970 and 1971 were built by the same female, or that the male who rested on exactly the same twigs below them in both years was the same individual. Probably these same two hummingbirds had occupied the slightly more distant sites in 1968. They appeared to be mated and the pair bond persisted from year to year. The relation between the sexes was much the same as in the Royal Flycatcher, in which the male closely attends the female and guards her nest but neither builds nor incubates nor feeds the nestlings (Skutch, 1960a).

It is sometimes roundly stated that hummingbirds of opposite sexes form no attachments that last beyond insemination and that the males fail to attend the nest. Nevertheless, Moore (1947) presented evidence that the males of certain species, especially the Sparkling Hummingbird of the Andes, help the female at the nest; and later, in the coastal range of Venezuela, Schäfer (1954) watched a male of this same species, distinguished by somewhat abnormal coloration, alternate on the eggs with the female and feed the young. Ruschi (1965), however, questioned the reality of male participation in nesting in the Sparkling Hummingbird. In Ecuador he watched a female incubate alone; and in his aviary two females reared five broods without help from a male at any stage.

Be that as it may, after attentively watching the Bronzy Hermits year after year I cannot doubt that the male and female were joined by a bond more enduring than that of all those hummingbirds in which the sexes separate after insemination. Throughout the incubation period, in three seasons, the male's association with the nest was even closer than that which we witness in many passerine birds, in which the male seems to lose interest during the monotonous days of incubation, although after the young hatch his interest revives and he brings food to them. Unlike certain other hermits and many other hummingbirds, Bronzy Hermits are not known to have courtship assemblies, which would hardly be compatible with the close association of one male with one female throughout the nesting period; but perhaps such assemblies have not been noticed because this species is so rare in this locality.

I am uncertain whether the situation in *Glaucis aenea* should be regarded as primitive in the hummingbird family or a new development which may some day lead to the male's active participation in parental offices. In any case, these observations should remind us how premature it is to make sweeping generalizations about the immense hummingbird family on the basis of the few dozen species that have been more or less carefully studied at the nest. Certainly some surprises await the diligent student of living hummingbirds.

The persistent aggressiveness of the second female at the first nest, in 1967, also calls for elucidation. The conspicuously hanging nests of hermit hummingbirds attract the attention of individuals other than their builders and owners more often than do the less obvious nests of other hummingbirds. A Band-tailed Barbthroat's nest was often inspected by a Long-tailed Hermit as well as by a Bronzy Hermit, and sometimes a barbthroat examined a Bronzy Hermit's nest. Less often hummingbirds other than hermits hovered in front of these pendent nests, as though interested in them. In another nest of the barbthroat two females laid, and both continued to visit it, although apparently all the effective incubation was done by one of them; the eggs failed to hatch. In Surinam the attention of Haverschmidt (1968) was drawn by repeated fights between two Rufous-breasted Hermits to a nest of this species beneath a fern frond. The nest contained three eggs which evidently were laid by two females.

It is not impossible that at my first nest of the Bronzy Hermit, which I did not find until it held two nestlings, two females had laid and one or two eggs had been thrown out by the strife between the two claimants of the nest, as had happened to one of the four eggs laid in the barbthroat's nest. Even on this supposition it is difficult to explain why the two Bronzy Hermits continued for nearly a month to contend for a nest with two young, when there was no lack of sites for other nests in the vicinity. In similar circumstances birds more sociable than hummingbirds would have cooperated more or less amicably in attending the nestlings, no matter whose they were.

I have never noticed any comparable behavior in other hummingbirds. In a garden in the Almirante Bay region of western Panama which in 1929 supported an extraordinary concentration of Rufous-tailed Hummingbirds, one often repeatedly visited another's nest to steal material for its own nest (Skutch, 1931). Aside from these instances, one hummingbird to a nest has been the rule in my experience with the family.

Summary

In a region where Bronzy Hermits are rare, a single female, possibly always the same individual, nested in or beside a small banana plantation in five consecutive years. Each nest was attached by cobweb beneath a narrow transverse strip of a great leaf of the banana or wild plantain, at heights of 12 to 20 feet. Constructed by the female of fine, dark, fibrous materials, the nest was a resilient open fabric through which much light passed.

Eggs were laid in the rainy season, from late May to early August. In one year, when breeding began exceptionally early, the female built a second nest close above her first and tried unsuccessfully to raise a second brood in this two-story structure.

The set always consisted of two tiny, elongate, white eggs which were incubated by the female alone. She sat rather inconstantly for 62 per cent of one morning and 58 per cent of another. The incubation period was about 17½ days.

Only the female fed the nestlings, bringing food at a rate which increased from 1.2 times per hour during their first few days to 1.7 or even 2.9 times per hour after the nestlings were older.

Before the nestlings could eject their droppings over the nest's rim, their mother carried them away in the tip of her bill.

The nestling period was at least 24 days. A young hummingbird who remained in the nest for 29 or 30 days and left only when disturbed had injured a wing by striking it against the banana leaf while taking the usual wing exercises. For successfully rearing the young a nest should be situated where the birds can exercise their wings without obstruction.

An unemployed female persistently persecuted a breeding female the whole time she was raising her nestlings, often interrupting feeding by her vicious attacks. Sometimes the two females grasped each other's bill.

Lasting pairs are formed. During the whole of one nesting, from building to the departure of the young, a male spent much time every day resting directly below the nest, at intervals flying up to examine it. Although he neither built, incubated, nor fed the nestlings, he guarded the nest against the intrusion of other hummingbirds, especially female Bronzy Hermits. In the following year, when the nest was in almost the same position, the male perched in the same coffee bush and behaved the same way. Aside from the disputed case of male attendance at the nest in the Sparkling Hummingbird, this appears to be the only reported instance of pair formation in the hummingbird family.

BLUE-CHESTED HUMMINGBIRD

Amazilia amabilis

The Blue-chested is a small hummingbird about 3½ inches long. In the male the forehead and crown are intense metallic green. The sides of the head and neck and the upper parts of the body are bright metallic bronze-green, becoming more bronzy on the rump and upper tail coverts. The two central tail feathers are dark metallic bronze or purplish bronze and the others are largely blackish, glossed with bronze or bluish. The remiges are dusky. The chin and upper throat are greenish or dusky. The lower throat and upper chest are metallic blue or violet-blue; the lower chest is dull green or bluish green; and the more posterior under parts are largely grayish. The female is similar but her colors are paler, with the blue of her chest spotty rather than continuous. The bill is straight and of moderate length, with a black upper mandible and the lower mandible reddish or yellowish with a dark tip. The eyes and feet are dark.

The Blue-chested Hummingbird ranges from Nicaragua to Colombia and western Ecuador. It prefers the more humid regions and in Costa Rica inhabits both the Caribbean lowlands and foothills and the southern part of the Pacific watershed, where it occurs abundantly from sea level up to about 2,500 feet and is occasionally seen as high as 4,500 feet. Avoiding the depths of closed forest, it frequents light open woodlands and thickets, dooryards with flowering trees and shrubs, coffee plantations, and similar areas. As in many other hummingbirds its seasonal movements are erratic and little understood, but evidently are controlled at least in part by the abundance of flowers. Suddenly it vanishes from a district where, a little while before, it was present in fair numbers, and then as suddenly it reappears. It visits a wide variety of flowers and is especially attracted to trees of *Inga* spp. (called "guava" in Costa Rica) when they are covered with masses of white, powder-puff blossoms.

Voice and Courtship

This is one of the many species of hummingbirds in which the males sing in assemblies, within sight and hearing of each other. Each can be found day after day on the same perch, a dead twig, the bare basal portion of a slender living branch, or a thin vine stretched horizontally between two trees—always where the performing bird is free of concealing foliage. These display perches are mostly between about eight and 20 feet above the ground, in tall but light second-growth woodland or in a shady garden. The singing males are usually separated by 50 to 100 feet. Four who performed amid the shade trees north of our house in 1961 were stationed at the corners of a trapezium with sides of 55, 55, 90, and 70 feet. One hummingbird sometimes performed only 40 feet from another, but this was evidently too close, for he was often to be found on an alternative perch 55 feet from this neighbor. Eighty feet from the northeastern angle of this trapezium was a fifth Blue-chested Hummingbird, easily recognized by his peculiar song, to which we shall presently return. Only 25 feet from the southeastern corner a Rufous-tailed Hummingbird had his singing post, and 50 feet beyond was another male of this species.

Although hummingbirds are among the most persistent songsters, few are melodious to our ears. The Blue-chested is no exception. His sharp, metallic *tsweet tswe we we we, tsweet tswe we we we . . . ,* interminably reiterated, is too thin and dry to please us. He sings as though he were in a feverish hurry. By contrast the performance of his frequent neighbor, the Rufous-tail, seems calm and leisurely, almost languid.

Many hummingbirds, including the Green Violet-ear, the White-eared, the Violet-headed, the Blue-throated Golden-tail, and the hermits (*Phaethornis*) sing with admirable persistence through much of the day. Others, such as the Rufous-tailed, restrict their sustained singing to the early morning until about sunrise. The song period of the Blue-chested Hummingbird varies in an interesting fashion with the intensity of his motivation. At certain times, as at the outset and end of the singing season, he performs only or chiefly in the morning from early dawn until about sunrise. At a somewhat higher intensity there are two song periods each day, in the early morning and again for perhaps an hour as the day ends. Between these two principal periods there may be a little sporadic singing, especially on dull, cloudy days. But when their zeal is greatest these hummingbirds pour forth their sharp, thin notes through much of the day, beginning in the dim dawn and continuing until the evening when there is hardly enough light to distinguish their tiny forms. Now even hours of hard rain will not silence them.

In the Valley of El General some hummingbirds, including the Violet-headed and the Blue-throated Golden-tail, sing in the dry season and fall silent as the rains become heavy. Others, including the Scaly-breasted and the Little Hermit, perform through most of the year, except in the driest weather. To this class belongs the Blue-chested Hummingbird. Although I have heard it singing more or less persistently in every month, it is likely to be silent for a while in February, March, and early April, especially if the dry season is severe. During the remainder of the year there are fluctuations in the volume and intensity of song that are difficult to explain.

I made many observations on the singing of these hummingbirds in the five years from 1958 to 1962, when they maintained a flourishing assembly among the shade and fruit trees north of our house. At the beginning of 1960, for example, they were singing much. Toward the end of January song waned and through February and March there was only a little sporadic singing, although these months were not severely dry. In April and early May they sang chiefly before sunrise. In the second half of May, which was a very rainy month, song increased and through June, July, and August much was heard in both the mornings and evenings. Late in September, which this year was drier than average, these hummingbirds were performing most of the day. Then singing waned again and in late October and through much of November it was largely confined to the early morning. With the advent of drier weather in December song increased once more, until by the solstice it continued through much of the day.

Through the early part of the following year, 1961, song diminished until by mid-March (which this year was very dry) it ceased completely. A good shower on 18 March was followed by increased singing during the remainder of the month; but in early April, when the days became oppressively dry, hot, and smoky, the hummingbirds fell silent. Increasing with the return of frequent rains in late April, singing continued at a fairly high intensity, but with puzzling minor fluctuations, through the remainder of this year. It was particularly profuse in the rainy months of October and November when it could be heard at almost any hour of the day.

In the mid-1960's, when I was absent from the farm for long intervals, this singing assembly was unaccountably abandoned although the area where it had been established had altered little.

The singing assemblies that I have observed have not been in places rich in flowers. Most of the hummingbirds proclaimed their presence persistently in spots where they had no sources of nectar to defend. The purpose of so much vocal effort is undoubtedly not to assert possession of a feeding territory but rather to advertise the male's availability to the other sex. Yet in the Blue-chested Hummingbird, as in other species that dwell among profuse tropical vegetation, I have never succeeded in witnessing the consummation of courtship. A second hummingbird, of undetermined sex, suddenly approaches the singing male; a swift pursuit carries the diminutive birds beyond view; and presently the songster is back again on his favorite perch, sounding his thin, metallic *tsweet tswe we we we* as hurriedly as ever.

Longevity of an Aberrant Songster

On certain evenings early in 1956 I heard the song of a Rufous-tailed Hummingbird. At first, busy with other things, I gave little attention to it, for Rufous-tailed Hummingbirds are common about our house and I had listened to their songs innumerable times. After it occurred to me that a Rufous-tailed Hummingbird should not be singing so late in the day, I went to investigate and saw that the author of this song looked far more like a Blue-chested Hummingbird than a Rufous-tailed. I surmised that he was one of those hybrids that are from time to time recorded among hummingbirds and that from one parent he had inherited the song of the Rufous-tailed, while from the other he had inherited the appearance and times for singing of the Blue-chested. Not only did this peculiar hummingbird persistently repeat the subdued little *tsewe tsawe* or *right up that tree* of *Amazilia tzacatl,* he also frequently uttered, with widely opened bill, a distinctive loud call which this species often emits, especially around midday in September and October.

This puzzling hummingbirds' singing post was close by the huge rock that rises beside the stream north of our house. Less than 50 yards away several Blue-chested Hummingbirds had their regular singing stations. Since I could discover no difference in plumage between this hummingbird and the others who sang typical Blue-chested songs, I finally concluded that he was no hybrid but an *Amazilia amabilis* of pure descent who had *learned,* rather than *inherited,* the song of an allied species common in the vicinity. That hummingbirds learn rather than inherit their little songs was also indicated by observations that I had made long before on White-eared Hummingbirds in the Guatemalan highlands (Skutch, *in* Bent, 1940:454-455). There were great contrasts between the voices of different individuals, yet as a rule the members of a single singing assembly sang very much alike. At first I was inclined to attribute these variations in voice to heritable differences in the structure of the hummingbirds' vocal organs; but as Ernst Mayr pointed out to me, these assemblies in the same district were not sufficiently isolated from each other to preserve consistent genetic differences. More probably young males learned the song type that prevailed in the assembly to which they attached themselves.

Year after year the Blue-chested Hummingbird sang his borrowed song in the trees above the big rock. As time passed I noticed that his tempo had become much faster than that of the Rufous-tailed Hummingbird, although in phrasing his song still closely resembled that of this species. He now sang in the hurried manner typical of the Blue-chested Hummingbird, but in an alien tongue. In 1961 I recorded his presence in the usual spot from March to August. Then after a long interval of silence, I again heard his song by the big rock on 29 January 1962. Now he started to perform later in the morning and sang much less than some neighboring Blue-chested Hummingbirds. Evidently his vigor was declining; yet he was still present at his old post at the beginning of September of the same year, when he sang in both the morning and the evening, although he did not continue so long as nearby hummingbirds of his kind. My last record of his presence was on 4 September 1962. By the middle of that month a Blue-chested Hummingbird with a normal song had established himself by the big rock and I never again heard the aberrant songster. Since he was at least a few months old when I first became aware of him in January 1956, he had survived for no less than seven years.

Information on the longevity of free hummingbirds is difficult to obtain. In the Bronx Zoo, a Green-throated Carib lived for 10½ years, and several other species in air-conditioned, light-controled exhibits survived for seven and eight years (Conway, 1967:8-9). A Glittering-bellied Emerald likewise lived for eight years in captivity (Bourliere, 1946).[1]

Nesting

The two completed nests of the Blue-chested Hummingbird that I have seen were in an orange and a grapefruit tree in the pasture in front of our house, at heights of six and 14 feet. Built upon slender, horizontal, leafy branches, they were neat open cups, in shape resembling a sphere with the upper third cut away. One of these nests, somewhat flared out by its occupants, measured 1⅞ inches in outside diameter by 1⅝ inches in height. The inside diameter was 1½ inches and the depth was ¾ inch. Externally it was composed of light-colored bast fibers and seed-down, with about half the surface covered by green lichens. The inner layer consisted largely of seed-down, with a few large, brown scales or ramenta from the stipes of tree ferns. The second nest had few lichens to cover the shredded bast fibers that gave a very light color to the outside.

On 15 January 1957 the first of these nests held two feathered nestlings, one of whom left before 07:00 on 17 January and the other between 07:00 and 07:20 on the following morning. On 14 February 1961 the second nest contained two nestlings becoming feathered, but a few days later they and their nest had inexplicably vanished. When I visited these nests the parent hummingbird rested upon a neighboring higher branch and chittered endlessly, complaining much as Rufous-tailed Hummingbirds do in similar circumstances.

In addition to these two completed nests I found females starting to build on 6 January and 18 May, but both of these undertakings were soon abandoned.

Summary

Blue-chested Hummingbirds frequent light, open woodland and thickets, dooryards with flowering trees and shrubs, coffee plantations, and similar areas. They suddenly appear and vanish from a locality; their movements are little understood.

Males join in courtship assemblies, in which each repeats his sharp, metallic notes persistently from an exposed perch, eight to 20 feet above the ground and 50 to 100 feet from the nearest competitor. They perform through most of the year, but if the dry season is severe and prolonged they are silent from late January until early April. The hour of singing varies with the intensity of motivation. At lowest intensity, as at the outset and end of the singing season, these hummingbirds perform only or chiefly before sunrise but at fullest inten-

[1] In a recent letter, Dr. Conway told me that a Purple-throated Carib lived for nine years and eight months after arriving at the New York Zoological Park.

sity they continue through much of the day. Year after year these assemblies occupy the same site, which is often poor in flowers.

One male Blue-chested Hummingbird persistently repeated a song that he had evidently learned from neighboring Rufous-tailed Hummingbirds, but he sang in the evening as Rufous-tailed Hummingbirds rarely do. He occupied the same singing post for at least six years and eight months, and must have been no less than seven years old when last heard.

In two nests in the Valley of El General eggs were laid in December and January, at the beginning of the dry season.

SNOWY-BREASTED HUMMINGBIRD

Amazilia edward

Slightly less than four inches long, the Snowy-breasted Hummingbird is metallic green on the crown, hindneck, and upper back; golden-bronze on the lower back, rump, upper tail coverts, and wing coverts; glittering metallic green on the face, throat, and upper breast; and on the more posterior under parts pure white. The wing and tail plumes are blackish. The nearly straight, moderately long bill is black, becoming red toward the base of the mandible. The sexes are similar. The foregoing description is of the race *niveoventer* of western Panama and Costa Rica, formerly considered a distinct species. Wetmore (1968) has recently discussed geographic variation in this hummingbird.

The Snowy-breasted Hummingbird ranges from eastern Panama to the southern Pacific quarter of Costa Rica. Altitudinally it extends from sea level up to about 5,500 feet on the Volcán de Chirirquí in extreme western Panama and to at least 5,000 feet in neighboring Costa Rican mountains. An inhabitant of flowery thickets, tall but rather open second-growth, and roadside trees, it is by no means so abundant and familiar in flower gardens, plantations, and open country generally as its sympatric congener, the far more widely ranging Rufous-tailed Hummingbird. It avoids the forest even more consistently than the latter.

Carriker (1910:528) was impressed by the abundance of Snowy-breasted Hummingbirds about the indian village of Boruca in the lower Térraba Valley. In 1907 he found it very common there, even though in the preceding year the insatiable collector C. F. Underwood had taken over one hundred specimens in this same locality! Nowhere have I found this pretty hummingbird more numerous than in the lower valley of the Río Buena Vista, around 3,000 feet above sea level on the northern side of the basin of El General. Here, in December 1935, it was common in thickets that flowered profusely as the drier weather began. It was especially attracted to the clustered florets with long-exserted white stamens of carboncillo (*Calliandra portoricensis*) and the similar, but red, flowers of *C. similis*. As the drought grew more severe in February and March and flowers became less abundant, these hummingbirds vanished, to return in smaller numbers after the rains had revived the vegetation in late April and May.

Here at Quizarrá, about six miles away in an airline and 500 feet lower, I have in later years noticed similar seasonal fluctuations in the abundance of Snowy-breasted Hummingbirds. They did not become numerous here until December 1943, two years after my arrival, when many were attracted to the long hedge of the verbenaceous shrub *Stachytarpheta mutabilis* that I had planted and was now profusely displaying its little purple florets. By the following March these hummingbirds had become rare. Each year, if the dry season is severe, hummingbirds of several kinds diminish in abundance or disappear toward its end, when wilting herbs and shrubs outside the forest flower more sparingly or not at all. I do not know where they go.

Like other hummingbirds, the Snowy-breasted visits a variety of flowers. In addition to *Calliandra, Stachytarpheta,* and the handsome rubiaceous shrub *Palicourea guianensis,* I have noticed its preference for the small spurred flowers which in April and May spread a golden mantle over the shapely rounded crowns of tall mayo trees (*Vochysia aurea* and *V. ferruginea*). One hummingbird to a tree seems to be the rule. Indeed, as far as I have seen, other kinds of hummingbirds are little attracted to these conspicuous flowers, which appear to be pollinated chiefly by bees.

Some years ago a Snowy-breasted Hummingbird claimed as its own a certain length of the *Stachytarpheta* hedge around our garden. This same hedge was frequented by two male White-necked Jacobin Hummingbirds, irregular visitors here, who evinced no antagonism toward each other and were so fearless that I touched one as he hovered close to me. Again and again the Snowy-breast darted truculently toward one of the bigger, blue, green, and white Jacobins. Sometimes the latter avoided contact by slipping deftly sideward in the air and sometimes I heard a sharp sound as the wings of the two flying birds struck together. Usually the Jacobin continued to make the rounds of the purple florets, taking little notice of his pestiferous assailant, but once he retaliated, driving the smaller hummingbird high into the air until pursuer and pursued looked like small black flies against the sky. After this sally the Jacobin returned to resume his visits to the flowers and suffer the continued aggression of the Snowy-breast. Yet I have seen the latter driven away by a Rufous-tailed Hummingbird, into whose feeding territory, a group of *Palicourea* shrubs, the Snowy-breast had intruded.

I have never found a singing assembly of Snowy-breasts nor noticed any spectacular aerial display. A solitary bird, probably a male, perches unobtrusively in a shade tree in a coffee plantation or in one by the roadside. At irregular intervals it repeats, in a small voice that always seems far away, a characteristic phrase which sounds like *d'be bee,* or *be bee bee.*

Nesting

Like a number of other hummingbirds of higher elevations in Central America, especially on the Pacific side, the Snowy-breasted begins to nest as the wet season passes into the dry, when the combination of moist soil and sunny sky favors the greatest profusion of flowers which supply abundant food for the nestlings and their parents. The earliest of the six nests that I have noticed in the Valley of El General, 2,400 to 3,000 feet above sea level, was started in mid-September 1967 when rains were hard and frequent. Before I could learn what, if anything, this nest contained, it vanished, perhaps having been knocked down by a fruit falling from the pejibaye palm (*Guilielma utilis*) beneath which it was situated. The next earliest nest was built in late October 1943 and eggs were probably laid in it about the end of that month, as on 8 December it contained two well-feathered nestlings that two days later had flown. In the other four nests eggs were laid in December and the nestlings were raised in January, when sunshine and flowers had become abundant. Three of these four nests were found at Rivas, in the valley of the Río Buena Vista, in January, 1937.

Hartman (1957) found a new but empty nest of this species near the village of Volcán, Province of Chiriquí, Panama at an altitude of about 4,500 feet on 14 March 1956. Unfortunately, he shot the hummingbird without waiting to see whether she would lay at this late date. Sometimes nests built late in the season remain empty.

The six nests that I have seen were situated from five to 30 feet above the ground; four of them were from five to 12 feet up, one at 18 feet. Two were in coffee shrubs, one in a tree growing in an open field, one on a roadside branch, the two highest in rather open spots amid tall second-growth woods. These nests rested in a fork or, more often, upon a thin or moderately thick horizontal branch beside an upright or ascending stem that gave lateral support. They were compact, downy, open cups, well covered on the outside with lichens and sometimes also green moss. Each of five nests held two eggs or nestlings. While I watched a parent attend two nestlings in a coffee bush, a second Snowy-breasted Hummingbird, evidently a male, perched high in a shade tree repeating from time to time his dreamy *d'be bee,* showing no interest in the nest.

Summary

The Snowy-breasted Hummingbird inhabits flowery thickets, plantations, light second-growth woods, and roadside trees, from sea level up to 5,000 feet or a little higher. Its abundance fluctuates seasonally.

Neither singing assemblies nor aerial displays have been noticed, but solitary birds, probably males, perch unobtrusively and utter at irregular intervals a characteristic phrase that sounds like *d'be bee* or *be bee bee.*

In the Valley of El General, 2,400 to 3,000 feet above sea level, this hummingbird nests chiefly in December and January, as the wet season passes into the dry and flowers become more abundant. Five nests contained the usual two eggs or nestlings. A nest was built in September in Costa Rica and one in March in Panama; eggs were not seen in them.

BLUE-THROATED GOLDENTAIL

Hylocharis eliciae

A hummingbird only 3½ inches long, the Blue-throated Goldentail has largely metallic green upper plumage, which becomes golden or coppery bronze on the upper tail coverts. Its tail is brilliant golden bronze, sometimes with a greenish tinge. Its throat and upper chest are bright metallic violet-blue; its breast, sides, and flanks metallic bronzy green; and its posterior under parts grayish or brownish. Its straight, light red bill is tipped with black. The female resembles the male but is slightly duller.

The Blue-throated Goldentail ranges from the Mexican states of Veracruz and Chiapas to Darién in eastern Panama. In northern Central America it is found on both slopes but in the south it occurs chiefly on the Pacific side, with a few straggling across the continental divide where it is low, as in the Canal Zone (Eisenmann, 1952:27) and northern Costa Rica (Slud, 1960:95). Altitudinally, it reaches upward to 3,600 feet in Honduras (Monroe, 1968:181). In Costa Rica it is, according to Carriker (1910:531), "scarce at very low altitudes, preferring an elevation between 1,000 and 2,000 feet, and is most abundant in the Térraba Valley." If one adds 500 feet, this agrees with my own experience. Although I worked at various points from 2,000 to 3,000 feet in the Térraba Valley between 1935 and 1940, I did not meet this hummingbird until I came to my present abode at about 2,500 feet beside the Río Peña Blanca below El Quizarrá in 1941. At first fairly abundant here, it has become rarer as the original forest has been replaced by pastures, coffee plantations, tobacco fields, and other cultivation. Yet in this region it is not an inhabitant of old forest so much as of light, moderately tall second-growth woodland and groves, from which it enters more open areas to visit flowers. In Panama, however, Wetmore (1968:319) knew the goldentail as "a forest inhabitant that seeks shaded haunts to perch, usually on the lower branches."

Courtship Assemblies

Early in 1942 I discovered two courtship assemblies of the goldentails. The first was situated near the Río Peña Blanca below our farm in an open stand of burío (*Heliocarpus*) and other swiftly growing trees beside a log-strewn field of maize. The three hummingbirds of this group perched on exposed, leafless twigs near the edge of the grove at heights of about 20 to 40 feet, where they would have been conspicuous had they not been so minuscle. Each repeated tirelessly a quaint, breezy little ditty that sounded like *see—sa-se se se se se* in an accelerated sequence.

The second assembly was also situated among second-growth trees, a mixed stand of half-grown targuá (*Croton draco*), guarumo (*Cecropia* sp.), guava (*Inga* sp.), and *Goethalsia meiantha* that formed a light, open grove on our farm between the high forest and the river. This gathering also consisted of three goldentails who perched on exposed, leafless, horizontal twigs from 15 to 40 feet above the ground. The two most distant of these performers were about 100 feet apart. This assembly, which never grew in membership, persisted in the same place until 1952, or for 10 years after its discovery. Probably it was abandoned because the *Goethalsia* trees, growing very tall and displacing most of the short-lived associated trees, made a heavier, more closed woods than goldentails choose.

The hummingbirds in this second group had a song quite different from that of the first group. I wrote it *zee w'w'w'w'w' zewét zewét zewét,* the part represented by the *w*'s being a sort of dry trill. The final disyllable, *zewét,* was repeated briskly a variable number of times. While proclaiming himself, each performer turned his head from side to side and at intervals rapidly shook his purplish black wings or spread his double-rounded golden tail. After singing a good while he would dart off beyond sight to search for food, as the immediate vicinity contained few flowers. Soon he returned to his favorite perch and resumed his song—and so he continued tirelessly to repeat the same phrase throughout a long day.

Although the birds in the two assemblies had quite different songs, members of the same group consistently repeated similar phrases. Since these two groups of birds were not reproductively isolated, it is unlikely that their song types were genetically determined. More probably, the hummingbirds in each assembly imitated its founder, or perhaps its dominant member, and the song type was preserved by tradition rather than by genetic transmission. In the assembly that persisted for ten years we must suppose either that one or more of its members was present throughout this interval—not an impossibility in view of the observed longevity of another hummingbird, the Blue-chested (see p. 49)—or that each season at least one bird from the preceding season returned to mark the location and carry on the vocal tradition.

Long ago in the congeneric White-eared Hummingbird in the Guatemalan highlands, I noticed a similar uniformity of song type within each assembly associated with striking diversity between assemblies. A somewhat similar situation has subsequently been described for the

Little Hermit by Snow (1968) and Wiley (1971). These observations strengthen the conclusion reached in the preceding chapter that among hummingbirds the song pattern is learned rather than inherited, as is true of a number of Oscines. Or if there be a basic hereditary pattern subject to modification by learning, as in the Chaffinch (Thorpe, 1958), the learned component far outweighs the innate foundation, at least in such hummingbirds as the Blue-throated Goldentail, the White-eared, and the Blue-chested. In view of the strong individualism of hummingbirds, their widespread failure to pair, and their lack of group impulses, this tendency to learn from their neighbors is surprising. The foregoing remarks refer to the louder, stereotyped songs that are tirelessly repeated in the singing assemblies. Hummingbirds of many species, some of which are not known to use advertising song, have a pleasantly varied whisper song, often so low that even at close range one first becomes aware that they are singing by noticing the vibrations of their distended throat. This is evidently innate and corresponds to the first rambling, nonspecific essays at singing of certain Oscines, notably the wrens.

The species of hummingbirds in which I have found singing assemblies include the Band-tailed Barbthroat, Green Hermit, Long-tailed Hermit, Little Hermit, Scaly-breasted Hummingbird, Rufous Sabrewing, Violet Sabrewing, Green Violet-ear, Violet-headed Hummingbird, White-eared Hummingbird, Blue-throated Goldentail, and the Blue-chested, Rufous-tailed, Amethyst-throated, and Wine-throated Hummingbirds. Sometimes one of these hummingbirds performs alone, perhaps because he has not yet attracted others; and solitary calling appears to be usual in the Snowy-breasted Hummingbird. In South America singing assemblies, or at least sustained singing, has been recorded for several additional species, including the King Hummingbird (Davis, 1958), *Leucippus chionogaster,* and three species of violet-ears (*Colibri*) (Niethammer, 1953:289-291). At the proper season in the high Andes the traveler with ears well attuned to nature's minor sounds soon becomes aware of violet-ears singing in almost every grove.

In eight of the 15 species of Central American hummingbirds known to have singing assemblies, the sexes are so similar in appearance that it is difficult or impossible to distinguish them in the field. In brilliance these eight species range from the dull brownish hermits to the glittering Green Violet-ear and Blue-throated Goldentail. In only four of 15 species with singing assemblies—the Violet Sabrewing, the White-eared, Amethyst-throated, and Wine-throated Hummingbirds—are the sexes strikingly different. On the other hand, in eight hummingbirds of the United States with spectacular aerial displays but apparently without singing assemblies, the females are very different from the males, lacking the glittering gorgets which the latter flash as they dash past. In other groups of birds a strong correlation between courtship and mating habits and plumage is evident: in Neotropical Oscines, for example, the sexes are much more often alike in constantly mated than in seasonally mated species. It appears that in hummingbirds singing assemblies tend to be associated with similar plumage in the two sexes, whereas more dynamic modes of courtship favor the development of contrasting plumage in the male and female. Wolf (1969) has argued that female hummingbirds who defend feeding territories have evolved plumage like the males' to give them equality with the latter in aggressive encounters. Effectiveness in territorial defense is probably only one of the factors favoring similarity in plumage between the sexes of hummingbirds, and possibly not the most effective. We need to know much more about the whole pattern of life of these birds before passing judgment on these difficult questions.

Vocal advertisement appears to be the prevailing mode of courtship among hummingbirds, at least in Central America. Formal aerial displays, such as have been described for a number of species of more open country in Mexico and the western United States (Wagner, 1954; Hamilton, 1965; etc.), are rarely seen among Central American hummingbirds. The most strenuous display that I have noticed here is the long, U-shaped dive of the Broad-tailed Hummingbird, which ranges from the mountains of western United States to those of Guatemala. Elaborate aerial displays, performed in a very much smaller air space, have been witnessed in the Little Hermit and the White-crested Coquette (Skutch, 1951, 1961a, 1964a). In a number of hummingbirds that I have seen year after year, including the Blue-crowned Woodnymph, Purple-crowned Fairy, and Long-billed Starthroat, I have noticed neither a singing assembly, nor sustained solitary calling, nor a flight display. How the males of these species attract the females remains to be discovered.

Over an interval of a decade at the goldentails' assembly among the *Goethalsia* trees, and less extensively at other assemblies, I recorded the dates when these hummingbirds performed. Singing begins in the latter part of the rainy season; often I heard the goldentails in September and once even at the end of August. When rains are very heavy in October or November they sing less or fall temporarily silent; but with the advent of drier weather, around the end of the year, they come into full song and perform from early morning to evening. January and February are months of great vocal activity in the assemblies. If the drought becomes severe and wilting plants open few flowers in late February and March, singing wanes or ceases. The return of refreshing rains in March or April stimulates all-day singing again, but song may diminish if, after a few good showers in March, the drought is renewed in early April. But in early or late May, a month of frequent afternoon deluges, the goldentails enter a long period of silence, which is continued through June, July, and August. Thus these hummingbirds sing more or less freely through eight or nine months of the year, with intervals of reduced activity during very wet weather in

its head rapidly from side to side, just as did a Scaly-breasted Hummingbird who roosted in a tree beside the house. At first pronounced, these head movements gradually decreased in amplitude until they ceased entirely. One evening when the starthroat arrived at its sleeping place at 17:52, they continued for three or four minutes; but on the darkly overcast evening of 2 March, when the hummingbird settled on its roost at 17:36, the head movements persisted for a quarter of an hour. When finally they died away the bird rested with its bill pointing forward and tilted slightly upward, as is usual with sleeping hummingbirds, whether on their nests or perched. After dark the starthroat's eyeshine in a flashlight's beam was surprisingly bright, like a brilliant point of light. Roosting on a nearly leafless slender twig at the top of the tree, the starthroat appeared to enjoy maximum security from terrestrial enemies at the price of full exposure to aerial predators. Evidently the former were a greater hazard. It is significant that nests of this hummingbird are likewise fully exposed.

Starthroats suck nectar from a variety of flowers but they are especially attracted to those with long corolla tubes, which their long bills are well fitted to probe. Such are the white flowers of the banana clustered beneath an uplifted dull red bract at the end of the dangling inflorescence. To reach these the hummingbirds must hover with body nearly vertical and bill inclined strongly upward. But, above all, starthroats favor those species of poró (*Erythrina*) in which the flowers have long, slender, tightly folded red standards that completely enclose the reduced wings and keel, as in *E. rubrinervia* and *E. berteroana.* Two small trees of the latter species growing in our dooryard have for many years attracted starthroats, which arrive in November or December when the trees begin to drop their trifoliolate leaves and put forth their earliest blossoms. The hummingbirds remain through the drier months of January and February when the nearly leafless poró trees are in full bloom, to disappear in March when few flowers remain and the spreading boughs are putting forth new foliage.

When a starthroat takes possession of a poró tree it repels all other hummingbirds of its own or other species. Of the local hummingbirds, only two in addition to the starthroat have bills fitted to reach the nectar of the tightly closed flowers. The first is the Scaly-breasted Hummingbird which has some difficulty inserting its fairly long bill into the sword-shaped red standard; when possible it does so from a convenient perch. The second is, oddly enough, a hummingbird with a very short bill, the Purple-crowned Fairy. Never attempting to reach the poró's nectar in the usual way, the Fairy hovers beside the flower and pushes its exceptionally sharp bill through the thick fleshy calyx that forms a collar around the base of the standard, making a tiny perforation through which it extracts the sweet fluid. The Scaly-breasted Hummingbird also frequently uses this method to reach the nectar.

With these smaller hummingbirds the starthroat does not have much trouble; they must sip their poró nectar when the starthroat who claims the tree is not looking. But if a second starthroat contests the claim, there is a different story to tell. One December, as the poró trees were beginning to flower profusely, I watched a fierce encounter between two claimants. While one starthroat perched in the tree the other darted rapidly back and forth a few inches above its head, flying a foot or so in each direction. Then the flying bird attacked the perching bird from below, seizing the other's head in its long bill. The other seized the bill of its assailant, who for a few seconds hung suspended by the two clasped bills. Then they separated and darted away.

Starthroats are silent birds. The only notes I have heard were low, dry monosyllables.

Nest and Eggs

Year after year, as sunnier days returned in December or January, I watched a starthroat collect cobweb from the walls of our house, then fly away so swiftly that I could never trace her course to the nest she was evidently building. At last, on 24 November 1967 I found a starthroat making a nest. Her site was 15 feet above the ground in a small dead tree, far advanced in decay, in the hillside pasture behind our house. The incipient nest was in a fully exposed situation on the upper side of a bend in a short horizontal stub of a branch, scarcely an inch in diameter, just beneath a somewhat thicker ascending stub. It was a low ring of fine, dark vegetable materials, with a few light-colored lichens already stuck to the outside, fastened to the branch with much cobweb.

Building proceeded slowly. The greatest number of visits to the nest which I counted in an hour was 15, between 09:18 and 10:18 on the morning the nest was found. Six of these visits with material were made in the first quarter-hour, five in the last quarter-hour. Usually the builder came with cobweb which, after settling in the nest, she spread over the outside with her long bill, sometimes reaching far down to touch the supporting branch. Continuing to sit she rotated to face in different directions; the bouncing motion of her body suggested that she was kneading the inside with her feet. The tuft of pure white feathers between her rump and the flanks on either side often gleamed in the morning sunshine. Frequently she sat shaping the nest for a minute or two but sometimes she remained for only a few seconds. Less often than cobweb, she brought a flake of lichen which she worked into the inside of the nest beside herself. She had the queer habit of flying from the nest to the end of one of the surrounding stubs and perching there briefly, then to another and another, sometimes resting for a few seconds on each of four different branches of the dead tree before she flew away. The builder was always alone, except once when another starthroat hovered over her while she sat in the nest. Then both flew rapidly away.

Six days after I found this nest in an early stage of construction, it seemed finished. It was an open bowl, broad and shallow, composed largely of mosses, liverworts (especially *Frullania*), and vegetable down bound together and to the branch with abundant cobweb. A few pale gray lichens were stuck on the outside. It looked like a small excrescence on the upper side of the stub with peeling bark.

During the eight days that intervened between the virtual completion of the nest and the laying of the first egg, the starthroat from time to time spread more cobweb over the nest. I found the first egg at 06:50 on 8 December. It was not covered during the two succeeding nights. The second egg was laid between 05:50 and 07:10 on 10 December, two days after the first. These eggs were elongate and pure white as in other hummingbirds.

About 100 yards down the hill from the site of this nest I found another on 8 February 1969, in the midst of the dry season. About 35 feet up in the top of a dead guava tree in a pasture, it was fastened to the upper side of a branch about an inch thick that was bent in the form of an inverted L, just above the angle in the branch. Although wholly exposed, the nest was not easy to detect because its brown

FIG. 7. Nest of Long-billed Starthroat in a typically exposed situation. Los Cusingos, Valley of El General, Costa Rica, December 1967.

color blended so well with that of the weathered wood and peeling bark that it appeared to be only an excrescence on the branch. It was impossible to see the contents of this nest but incubation appeared already to have begun. A week later the nest was destroyed by some unknown agent.

In November 1964 Rowley (1966:152) discovered two nests of the Long-billed Starthroat in Oaxaca, Mexico, at an altitude of about 2,400 feet. In site and construction these nests differed greatly. The first was firmly attached by cobweb to the upper side of a bend in a small outer branch of a tree of *Cecropia peltata,* 50 feet up. Except that it was much higher above the ground and apparently more deeply cupped, it resembled the Costa Rican nests in site and construction (see Rowley, *op. cit.,* fig. 26). The second Mexican nest was built about eight feet up on a downward-sloping branch of a shrub growing in light shade in a coffee plantation. More loosely constructed than the first, it had many pieces of bark decorating the outside. On 19 November it held two fresh eggs that measured 12.8 by 8.8 and 13.6 by 8.5 mm. The first of Rowley's nests had apparently contained eggs on 6 November.

Incubation

I watched the starthroat incubate in the nest behind our house throughout the morning of 18 December and the afternoon of 22 December. She first left the nest at 05:25 in the drizzly dawn of 18 December and returned for the night at 17:18 on the clear evening of 22 December. I timed in full 15 sessions that ranged from three to 137 minutes and averaged 34.3 minutes. Her two longest sessions, of 137 and 67 minutes, were taken in the afternoon; her next longest sessions, 53 and 41 minutes, came in the late forenoon. Her 16 recesses ranged from four to 19 minutes in length and averaged 10 minutes. She incubated for 77.4 per cent of the day.

I did not see the starthroat turn her eggs with her long bill. Sometimes while sitting she bounced rapidly up and down, as though kneading the nest material or perhaps turning the eggs with her feet. In the hot sunshine of the early afternoon she usually sat facing away from the sun. Once, facing west, she leaned far over away from the sun, the feathers of her breast puffed out as though sunbathing on the nest. Rarely she panted with her neck stretched up. To leave the nest she flew upward and backward out of the bowl, then reversed to fly away. Once, while she sat, another starthroat hovered above her, then both darted away together. Otherwise no second hummingbird showed interest in the nest.

Four times during the forenoon the starthroat added material, usually cobweb, to her nest, making three special trips to bring it. The fourth time she stayed to incubate. From 16:56 to 17:03 she came to the nest with cobweb five times in seven minutes. Each time she sat in the nest to spread the cobweb over the outside, then flew away. Otherwise nothing was brought to the nest in the afternoon.

The first egg hatched in the afternoon of 28 December and by 07.00 next morning the second had hatched and all the shells had been removed. The incubation period was between 18½ and 19 days.

Nestlings

Development.—The dead tree on the steep slope was so far advanced in decay that I did not attempt to reach the nest it held, but made all inspections by means of a mirror raised on a long wild cane. At first I noticed no peculiarity in the nestlings' appearance. When they were two weeks old, however, their upper parts, throat, and breast were covered with short, whitish down, more abundant than that on any other hummingbird that I have seen. This blended with the light-colored bark of the dead branches on and amid which the nest was situated and made the nestlings less conspicuous in their exposed position. When they were 19 days old their remiges were becoming long. Dull black in color, these feathers contrasted with the whitish down that covered much of the body, as in young Brown Boobies. Now from time to time a nestling rose up and flapped its wings vigorously. They often shifted their position in the nest, preened much, and scratched their heads over the wing as adult hummingbirds do. When 23 days old these young starthroats were well feathered in brown and whitish body plumage. Now they sometimes preened each other as well as themselves. Two and three days later they left the nest.

Brooding.—While brooding small nestlings the parent sometimes bounced up and down, much as she had done while building—a puzzling activity at this stage. On the sunny morning of 5 January the two nearly naked nestlings were brooded for a total of 97 minutes during 391 minutes of observation, or one quarter of the time. Intervals of exposure were long, ranging from 21 to 61 minutes, and with a single exception lasting half an hour or more. The longest periods of brooding came not in the cool early morning but toward midday when the nestlings were exposed to strong sunlight. Early in the morning the longest interval of brooding was only 13 minutes, but in the late forenoon they were covered for 28 and then for 34 minutes, obviously to shield them from too-strong insolation. While the parent sat over them, the nestlings stuck their bills, or even their whole heads, out in front of her breast and panted. Despite some parental shading the young were often left exposed to full sunshine for intervals that would have been fatal to many a larger passerine nestling with more developed plumage. Nestling hummingbirds are marvellously resistant to heat, cold, and strong sunshine.

At night these nestlings were brooded until the elder of them was 10 days old, the parent sitting over them with her bill directed forward. After that they always slept alone, although at an altitude of 2,500 feet nights in January were often chilly. Such early cessation of brooding is usual in hummingbirds, except those at high altitudes where nights are frosty. At a neighboring nest of the Rufous-tailed Hummingbird the nestlings were not brooded at night after they were nine days old.

Feeding.—On 5 January, when the two nestlings were seven days old the parent came with food eight times between 05:38 and 12:09. On each visit she usually regurgitated from three to five times to each of the nestlings, alternately, thus delivering the contents of her crop in from seven to nine installments, although on one visit there were only three installments. After a generous meal the nestlings' necks were grotesquely swollen with food where the crammed oesophagal pouch protruded to one side.

On 17 January, when the two nestlings were 19 days old, the parent came with food 13 times between 05:30 and 12:00. Now she delivered it in only two to four installments, regurgitating to each of the young only once or twice on each visit. Probably more nourishment was passed at each act of regurgitation than when the nestlings were smal-

TABLE 1
Rates of Feeding Nestling Hummingbirds

Species	Hours watched	Feedings per hour: 1-4 days	5-10 days	11-15 days	16-23 days
Band-tailed Barbthroat *Threnetes ruckeri*	18½	1.3		1.5	2
Bronzy Hermit *Glaucis aenea*	12	1.2		1.7	
	6	1.3			
Long-tailed Hermit *Phaethornis superciliosus*	7			1.6	
Little Hermit *Phaethornis longuemareus*	12	1.3		2.5*	
	6				2.3
Scaly-breasted Hummingbird *Phaeochroa cuvierii*	18¼		2.3		2.5
Green Violet-ear *Colibri thalassinus*	8		2.6		
White-crested Coquette *Paphosia adorabilis*	15	2.4	1.8	2.4	
Purple-throated Mountain-gem *Lampornis calolaema*	5		2.4		
White-eared Hummingbird *Hylocharis leucotis*	13	1	1.8		2.5*
Long-billed Starthroat *Heliomaster longirostris*	13		1.2		2

* A single nestling was present. In every other observation period two were present; accordingly, to find the rate per nestling/hour, the figures should be halved.

ler. To deliver food the parent inserted her long bill far down into the nestlings' upstretched necks. Like other nestlings that are fed by regurgitation, hummingbirds receive food less frequently than many small birds that are fed directly from the parent's mouth (Table 1).

Departure.—At 07:15 on 23 January both young hummingbirds were still in their nest, but an hour later one was perching about two feet away. There was still much whitish down on the fledgling's forehead, center of back, throat, and flanks. Its tail was blackish with the three outer feathers on each side prominently tipped with white, most broadly on the outermost. Presently the mother came and regurgitated once to the fledgling. She offered it more, touching its bill with hers, but it did not respond so she went to the nest and fed the nestling who remained there. Returning to the fledgling outside she again offered to feed it but again it refused. The one in the nest flapped its wings long and vigorously, stretching up its legs high while clinging. At 08:27 the fledgling outside flew off beyond view. Ten minutes later the parent fed the stay-at-home in three installments. She went off, then returned and persistently offered it more but it was not hungry. This young hummingbird flew away between 13:30 and 15:10 in the afternoon of the same day, leaving the nest well flared out all around. The nestling period was 25 to 26 days.

Young After Leaving The Nest

Eight days after their departure the young starthroats could hover expertly and seemed to pluck things, too small to be distinguished by me, from the surface of fruits and twigs of an *Inga* tree close by the nest tree. Two days later I found both of them in the poró tree behind the house. They hovered in front of the red flowers and touched them with their bills, but they did not insert their bills into the corollas. Again they seemed to gather minute objects from foliage and twigs. On the morning of 3 February I saw both fledglings alight near their mother, eager for a meal; but one flew at the other and struck it with its body. After some skirmishing one flew away and only one was fed. After a while one was fed again. Later I saw one fledgling threaten and chase the other. Although only 36 days old and still dependent on their parent, they had already developed that intolerance of other individuals of their kind which is so highly characteristic of starthroats.

The parent continued to give some food to the juveniles, or one of them, until 15 February when it was 48 days old and had been out of the nest for 23 days. During the second week of February the young hummingbirds spent much time in a tree of *Cassia spectabilis* down the slope from the nest tree. On the morning of 7 February both juveniles rested peacefully a yard apart in this tree, preening and occasionally stretching, for many minutes. But when at last their mother came to feed them and the two alighted on opposite sides of her, one chased the other away; only one was fed.

An adult male starthroat, as I judged from the splendor of his magenta gorget, also spent much time in this *Cassia* tree and sometimes tried to drive away the other starthroats. When he assailed one of the perching juveniles, it hung beneath the twig and presented its bill to the attacker, who promptly flew away leaving the young bird in peace. Such clinging beneath a twig was a display of agility surprising in a weak-footed hummingbird. Thereafter the juvenile bravely stood its ground even when the adult male almost struck it. Sometimes there were swift pursuits in which three starthroats participated. A Snowy-breasted Hummingbird and two Scaly-breasted Hummingbirds also spent some time in the *Cassia* tree pursuing each other or being chased by a starthroat.

By mid-February when I last saw a juvenile receive food from its mother, it had a white patch on its rump, metallic blue feathers on its crown, and magenta feathers on its throat. It was already difficult to distinguish from her.

Summary

The Long-billed Starthroat inhabits semiopen country, including gardens, plantations, and light woodland. It is a solitary bird.

A starthroat roosted on an exposed twig at the very top of a dying tree. After settling on its roost in the evening it twitched its head rapidly from side to side, sometimes continuing for a quarter of an hour before it assumed its sleeping posture with bill pointing forward and upward.

Starthroats are especially fond of the long, red flowers of the poró (*Erythrina* spp.). When one takes possession of a poró tree it chases away all other hummingbirds.

Four records from southern Mexico and Costa Rica indicate a breeding season extending from November to February, in the early part of the dry season. The nest, a shallow bowl, is often in a very exposed situation.

The first egg was laid eight days after a nest seemed finished. The second egg was laid two days after the first, in the early morning. Only the female incubated. The incubation period was 18½ to 19 days.

As the nestlings grew older they became covered with short tufts of whitish down, unusually abundant for a hummingbird, that assimilated them to the light-colored bark of the surrounding dead branches.

When the nestlings were a week old they were brooded chiefly to shield them from strong sunlight. Nevertheless, they were left exposed to intense insolation for intervals that would have been fatal to many passerine nestlings. Nocturnal brooding ceased when the nestlings were only 10 days old, with feathers still ensheathed.

When the nestlings were seven days old they were fed eight times in 6½ hours of the morning; when 19 days old, they were fed 13 times in an equal interval, always by their mother.

The nestlings left the nest when they were 25 and 26 days old and could fly well. Although they soon tried to feed themselves they continued to be nourished by their mother for at least 23 days after they left the nest, or until they were 48 days old. Long before they became independent the two juveniles became antagonistic to each other, and one tried to prevent its brood-mate from being fed.

Family TROGONIDAE

MASSENA TROGON

Trogon massena

One of the largest and heaviest of the New World trogons, the Massena or Slaty-tailed Trogon is a stout bird about 12½ inches long. The male is largely glittering metallic green with a golden luster on the back and often a blue or violet iridescence on the lower rump and upper tail coverts. The sides of the head, chin, and throat are dull black. Behind the green chest the under parts are bright red. From above the tail is metallic green, tipped with black; from below it is slate-colored. The wings are largely black with the coverts and secondaries finely vermiculated with black and white, appearing light gray at a distance. The short thick bill, with coarsely serrated cutting edges on both mandibles, is orange-red. Around the dark eyes are pale red bare orbital rings. The legs and toes are gray. The female, as large as the male, is mostly slate-colored except for the abdomen and under tail coverts, which are red. Her upper mandible is black except at the base where it is orange like the lower mandible. Her eyes and feet are colored as in the male.

The Massena Trogon inhabits the wetter forests from southern Mexico to Ecuador and from sea level to about the upper limit of the Tropical Zone. At an altitude of nearly 4,000 feet on the Pacific slope of southern Costa Rica I found it abundant. It is fairly common throughout the Caribbean rainforests of Central America. From the heavy forest, which is its true home, it sometimes ventures a short distance into adjoining secondary woodland, plantations, or clearings with scattered trees to forage or even to seek sites for its nests. It generally remains high in the trees but may on occasion descend low for food or to carve its nest cavity. It perches with the dignified erectness characteristic of its family, its tail hanging vertically or even inclined slightly forward. Its flight, rarely long-continued, is as strongly undulatory as that of other trogons. An essentially solitary bird, it is rarely seen with others of its kind except at an attractive fruiting tree or when several gather to sing or call competitively among the forest trees in the early months of the year when mates are sought.

This is the only trogon that I ever found roosting. At La Selva for nearly a month in April and May of 1967, I kept under observation a male who slept nightly at the edge of the forest, beside an old banana plantation becoming choked with weeds. His roost was a densely foliaged branch leaning far out over the clearing at a height of about 15 feet. Soon after sunset he would appear in the top of a tall tree at the forest's edge above the place where he slept. After resting here a while he flew down to a lower perch and paused here, too. Then he dropped abruptly down to the small tree in which he roosted. He perched a few minutes on a rather exposed branch, then moved into the clustered foliage amid which he slept, always alone. His time for retiring was 17:50 to 18:00.

Food

Like related species, the Massena Trogon subsists largely upon small fruits which it plucks on the wing. Flying up to a berry or arillate seed, the bird seizes it in its strong serrated bill and either detaches it by its momentum as it swoops past or, if the fruit is more firmly attached, it pauses in the air and throws its weight upon the object in its bill to pull it free. Small and middle-sized palm fruits contribute largely to this trogon's diet. After digesting the soft outer layer the seeds are regurgitated, as are those of *Virola* spp. and other trees of the nutmeg family which are swallowed for the spicy red aril that embraces them, the seeds themselves being indigestible. Chapman (1929:53) recorded the fondness of this and other trogons for the small, hard berries of the mangabé tree (*Didymopanax morototoni*). The diet of fruits is varied by numerous caterpillars and mature insects which are snatched from foliage at the end of an outward or upward dart, without alighting, much as berries are picked from trees. If the victim is large the trogon beats it against its perch, first on one side and then on the other, throwing its whole body into the blow. Sometimes a caterpillar is plucked along with the leaf to which it clings and the latter is dropped as the bird swallows its meal. Once I saw a female Massena Trogon catch a gray lizard a few inches long and gulp it down without difficulty.

On Barro Colorado Island Stott and Selsor (1961) watched four adult Massena Trogons catching insects stirred up by a troupe of five White-faced Monkeys (*Cebus capucinus*) as they moved through the treetops. They were using the monkeys as anis, Cattle Egrets, and other birds use grazing animals to make their insect food more easily available, and as Bicolored Antbirds sometimes use men.

Voice

The call or song of the Massena Trogon is a deep, full-toned *wuk, wuk, wuk.* The regularly-spaced unmelodious notes are often repeated many times over. Eisenmann (1952:28) described this call as "a loud *cah, cah, cah,* interminably repeated at the rate of about two a second." This rather harsh utterance contrasts with the soft mellow *cow* notes of such trogons as the Black-throated, the Violaceous, and the Collared; it is never accelerated into a roll like the melodious call of the White-tailed Trogon. This *wuk, wuk, wuk* sounds through the Central American lowland forests chiefly during the drier months early in the year, but in El General I have heard it sparingly as late as early July.

Another utterance consists of full throaty notes, rapidly repeated. At times this sequence is accelerated into a long-drawn, rather wooden rattle, which is one of the frequent sounds of Central American warm forests. At other times the trogon's notes are best described as rapid cackles.

Nest and Eggs

In Central America the Massena Trogons begin to excavate their nest cavities in early March and continue their reproductive activities until June or July. As in a number of other trogons whose chambers are deep and well enclosed, these cavities may be carved into either decaying wood or large, hard, blackish arboreal termitaries. Of the five completed nests that I have seen, four were in very rotten massive stubs or stumps at heights of 8½ to 18 feet and one was in a termitary 15 feet up; but I have seen Massena Trogons working at cavities that were never completed in both a termitary and a trunk at heights of 40 to 50 feet. A thick, badly decayed trunk which one year contained an active nest 18 feet up contained one three feet lower the following year, the top of the stub having meanwhile decayed away. The nest in a termitary was beside a mountain torrent in the forested foothills of northern Honduras. Of the nests in trunks, two were within the forest, one was a few yards from the forest's edge in a newly burnt clearing planted with maize, and the lowest was 60 feet from the forest in a charred stump in a pasture where many young guava trees were growing up.

On 13 May 1954 I watched a pair of Massena Trogons excavating a cavity about 50 feet up in a massive trunk that was far advanced in decay and already contained many holes made by woodpeckers or trogons. The male and female went alternately to cling to the depression in the side of the trunk where she stayed for three or four minutes at a time, but he for only a minute or so. Clearly the female was taking the leading part in the task but even she seemed to accomplish little, to judge by the slight amount of debris that fell while her head was in the excavation. While work was in progress low, soft monosyllables floated down from the trogons almost continuously. This excavation was never used. When I tapped on the rotten trunk some days later, a Bushy-tailed Olingo emerged from one of the holes at the top, but it soon returned inside.

Whether carved in a trunk or a termitary, the nest cavities are similar in form and size. A completed chamber in a trunk was 8½ inches high and 6½ inches wide. It was entered through an obliquely ascending tube about 3½ inches in diameter and 5 inches long, which reached the rounded top of the chamber. A ridge of wood separated the bottom of the chamber, where the eggs rested, from the lower side of the entrance tube and prevented the eggs from rolling out. These nest chambers are never lined. At first the eggs rest upon a

FIG. 8. Termitary in which Massena Trogons nested. The boy's hand is in the doorway. Lancetilla Valley, Honduras, 13 June 1930.

litter of coarse fragments of decaying wood or of hard black chips from the termitary. As incubation proceeds, regurgitated seeds of various sizes and shapes accumulate on the bottom of the chamber and form a pebbly bed beneath the eggs. In contrast to the eggs in the shallow cavities carved in trunks by Black-throated, Collared, and Mountain (=Mexican) Trogons, which are readily visible from in front, those in the well-closed chambers of the Massena Trogon can be viewed only by inserting a mirror and an electric bulb through the entrance tube.

Nests with eggs were discovered in Caribbean Honduras on 7 June 1930 and in the Valley of El General in Costa Rica on 26 April 1944 and 7 May 1956. Each of these sets consisted of three eggs. A nest found in El General on 10 April 1959 contained two newly hatched nestlings and one pipped egg. The glossless eggs vary in color from palest blue to white. The measurements of nine eggs averaged 36.3 by 28.1 mm. Those showing the four extremes measured *38.1* by 28.6, 34.9 by *29.0,* and *33.7* by *27.0* mm. Those of a single set differed considerably not only in size but in shape, the thinner end of some being fairly pointed while on others this end was quite rounded.

The eggs in the termitary in Honduras had a speckled aspect caused by minute fragments of feathers, mostly gray but a few red, that were stuck to the shells by a gummy substance that had evidently been secreted by the soldier termites. So firmly attached were these bits of feathers that, until I succeeded in rubbing some off after wetting them, I was not certain whether these first trogon's eggs that I ever saw were mottled or not. In addition to the speckles, a whole side of one of these eggs was discolored by a dirty brown substance that evidently came from the termites' nest. When I removed these eggs for inspection a few soldier and worker termites crawled over my hand. These trogons had carved their chamber into an occupied termitary, as I once watched a pair of Citreoline Trogons do (Skutch, 1948).

Incubation

The nest in the termitary beside the rushing mountain stream in the Honduran foothills was the first nest of any kind of trogon that I ever saw, and indeed almost the first nest of a bird of tropical forest that I attempted to study. Occasional visits, when I tapped on the trunk that supported the termitary to make the incubating bird appear, disclosed that the male was on the eggs between about 08:00 and 16:30 and that the female was there in the early morning, the late afternoon, and presumably through the night. Provided with this information I decided to watch the changeover. Arriving at the stream in the late afternoon, I hid behind a ledge of rock upon which I had piled some leafy boughs through which I could peer at the termitary. At 16:34 the female trogon dropped down from the high trees overhead and alighted on a branch above the stream. Then she flew to a limb directly in front of the entrance. The male thereupon stuck forth his head and a few seconds later flew silently out and away. A minute later the female entered the black hole in the side of the termitary. The changeover was effected without a note audible to me nor any sign of greeting between the two partners.

At 07:15 next morning I returned to my observation post behind the ledge of rock. As I waited quietly beside the rushing water a Northern Royal Flycatcher fed her nestling in a yard-long nest hanging above the channel, to the accompaniment of wild piping notes from her and her mate. Finally, at 08:38 the male trogon suddenly alighted on his usual perch over the water and remained there, slowly raising and depressing his long green tail through a wide arc, while with grave deliberation he turned his head from side to side. Presently he flew to a second twig, then to a perch directly in front of the opening in the side of the termitary. Since his mate did not come out he called in a very low voice. This I knew because I saw his throat swelling out, although his notes were not audible above the clamor of the stream. He hovered for an instant before the doorway, then alighted nearby and called somewhat more loudly. Still receiving no response, he finally entered the nest (at 08:45) without awaiting the departure of his mate. Seven minutes later a trogon shot out of the termitary and flew off swiftly through the forest on the other side of the stream. In the dense shade its plumage looked so dull that I jotted in my notebook "8:52 F left," then crossed the stream to verify my statement.

To my surprise, no trogon flew from the termitary as I climbed a ladder set against the trunk. Unable to see inside when I reached the doorway, I returned to the ground and struck the trunk with a stone. Still no response! Taking my flashlight, I ascended again for a more thorough investigation. This time I noticed a number of downy gray feathers, some of which were tipped with red, on the topmost round of the ladder and on a large leaf of an epiphytic aroid close below it. I sensed tragedy. Cautiously I stuck my hand into the dark hole. My fingers encountered only fragments of eggshell, a few seeds regurgitated by the sitting birds, and particles of the termitary that covered the rough floor. Undoubtedly the trogon I had seen fly off in silence through the forest was the male who had entered only a few minutes earlier. Some animal had plundered the nest since his departure on the preceding afternoon, and I wondered whether the female had escaped with only the loss of a few feathers. Doubtless I had done wrong to leave the heavy ladder against the nest trunk between my visits of inspection; but in later years many another nest similarly situated has met the same unhappy fate although no ladder was near it.

Returning to this termitary later in the year I found that the termites had closed the hole in which the trogons nested.

After the loss of this first nest I searched long but vainly through the neighboring forest for another like it. Twenty-five years after this episode we made many observations, including one all-day watch, on the pattern of incubation at an inaccessible nest at Los Cusingos. The division of time on the eggs by the two sexes was essentially the same as at that first nest in Honduras, but the changeovers were not so silently effected. On 10 May 1955 I began to watch at 07:10, when the forest was dark and misty. By 08:30 the sun was sending its rays down through the treetops. At 08:43 a Massena Trogon called *wuc, wuc, wuc* in the distance. Ten minutes later he alighted on a vine in front of and well above the nest, where he delivered the long throaty rattle about 27 times. The female slid slowly out of the cavity, pausing long with her head in the doorway, then advancing gradually until almost all her tail was visible, then suddenly darting forward and away among the trees. Her emergence took about three minutes. The male then entered without further delay but the tip of his tail remained visible for some minutes before it finally vanished. Soon his red bill appeared dimly at the head of the entrance tube and he spent some minutes looking out. After a while he settled down inside and became invisible.

While watch was kept, the male remained in the hole for the rest of the morning and nearly all afternoon. Only once during this long session did his red bill become dimly visible in the entrance tube. For the first $2\frac{1}{2}$ hours of the afternoon rain came down in a steady deluge, to the accompaniment of peals of distant thunder. After the storm ceased the dripping woodland remained dark and gloomy until nightfall. At 17:00 the female suddenly appeared on a bough above and in front of the nest and delivered the long, throaty, rolling call eight times, while the male slowly emerged. After eight hours and five minutes of continuous incubation, he flew away. The female advanced to the frond of a chonta palm (*Iriartea* sp.) close in front of the doorway where she repeated the roll 11 times more, each time raising and lowering her tail. At 17:03 she entered and I left her there for the night.

On 8 May the male of this nest came to take his turn on the eggs at 08:27; on 11 May at 08:45. On 12 May he was still on the eggs at 16:35 in the afternoon. The Massena Trogon's pattern of incubation with only two changeovers each day, the male sitting continuously for many daytime hours, is essentially the same as that of a number of other lowland trogons including the White-tailed, the Citreoline, the Black-throated, and the Collared. In the highland Quetzal, however, the male takes two sessions each day, separated by one session by the female.

Of the five occupied nests of the Massena Trogon that I have found, four were known to have been pillaged before or soon after the eggs hatched. The fifth was inaccessible and its outcome was not learned. Accordingly, for an account of the care and development of nestling trogons I refer the reader to the study by Gross (1930) or to my papers on related species (Skutch, 1942, 1944, 1959, 1962c).

Summary

The Massena Trogon inhabits humid forests from sea level up to about 4,000 feet. It is generally seen singly, high in trees.

A male roosted alone, night after night, on a low, densely foliaged branch projecting from the forest into an abandoned banana plantation.

This trogon subsists upon a variety of fruits and arillate seeds, larval and mature insects, and rarely small lizards. It seizes the fruit, or catches the insect, while flying past or hovering on wing.

The call or song is a long sequence of loud, rather harsh, regularly spaced notes, easily distinguished from the softer and mellower tones of most of the trogons with which this species associates. It also utters a long-drawn, wooden rattle.

In Central America the breeding season extends from March to June or July. The nest, a well-enclosed chamber entered through an upwardly directed tube, is carved by both sexes into either decaying trunks or occupied termitaries at heights of about eight to 50 feet. Sometimes this trogon nests in a clearing near the forest.

The set usually consists of three eggs, unmarked white or palest blue. They rest on the unlined floor of the chamber or often on a pebbly bed of seeds regurgitated by the incubating trogons.

Both sexes incubate. Each day the male takes one long session, extending from the middle of the morning to past the middle of the afternoon and continuing for about eight hours. The female relieves the male in the afternoon and sits continuously until he returns on the following morning.

Four accessible nests were pillaged before, or soon after, the eggs hatched.

VIOLACEOUS TROGON

Trogon violaceus

One of the smaller members of the family, the Violaceous Trogon is about nine inches long. In the male the head and throat are black, merging into violet-blue on the hindneck and chest; remaining upper parts brilliant metallic green, more yellowish on the back, more bluish on the rump and upper tail coverts; central rectrices bluish green or blue, tipped with black; outer tail feathers narrowly and prominently barred with black and white; wing coverts finely vermiculated with black and white, appearing gray at a distance; remiges largely black; under parts, posterior to the chest, orange-yellow. His short, thick bill is gray and a prominent yellow rim surrounds his brown eyes. The female is largely slate-colored, with yellow under parts and barred outer tail feathers, much as in the male. She has a prominent white crescent behind each dark eye and a smaller one in front. From the Black-throated Trogon, which mingles with this species in wet lowland forests, the male Violaceous Trogon is readily distinguished by his golden rather than bluish white eye ring and his song, which is described beyond. Female Black-throated Trogons are brown rather than slaty. In the larger Citreoline Trogon, which the Violaceous meets in drier regions, the more or less extensively white-tipped black outer tail feathers of both sexes are not barred; the notes are quite different.

The Violaceous Trogon ranges from southern Mexico to Peru, the Guianas, and Amazonia. Although most abundant in the humid lowlands, in Central America from Panama to Guatemala, it extends upward to about 4,000 or 4,500 feet, and I find no record of its occurrence at higher altitudes in other parts of its extensive range. Although Land (1970:171) gives its upper limit in Guatemala as 1,850 meters or 6,000 feet, this is so much at variance with all other published records that I have seen, as well as my own experience, that it requires confirmation.

Violaceous Trogons are most often seen at the forest's edge, in thinned woodland or more open country with tall trees not too widely scattered. They frequent the shade trees of cacao and coffee plantations. In dense forest they tend to remain in the treetops and are more often heard than seen. They are found in drier woodland, such as that of much of the Pacific side of Central America, as well as amid the rainforests of the Caribbean slope. Along the coast they inhabit tall mangroves. Like other trogons they are essentially solitary birds, although a number may gather at an attractive fruit tree or to compete for mates as the nesting season approaches. They perch quite upright, with tail hanging almost vertical, and fly with intermittent wing beats and undulating course like other trogons.

Once I watched a male Violaceous Trogon sun himself on a high exposed, horizontal limb of an *Inga* tree in a coffee plantation. He held his tail almost horizontal and turned to one side, with all the feathers fanned out, the barred outer rectrices conspicuous. His wing feathers were spread over his flanks below his tail with the yellow plumage showing between their separated ends. He scratched himself and preened; to scratch his head, he raised his foot inside and over his drooped wing, in the manner of songbirds, as I repeatedly saw.

The food of this trogon seems to differ little from that of other members of the family. They pluck a variety of fruits and catch insects in the spectacular manner described for the Massena Trogon. I have seen them tear pieces from the dangling fruiting spikes of *Cecropia* trees in this fashion, carrying each larger fragment to a perch before swallowing it. Chapman (1929:53) watched these and other trogons on Barro Colorado Island eating the small berries of the mangabé tree (*Didymopanax morototoni*), and he saw "yellow-bellied trogons," doubtless the present species, picking something from wasps' nests. We shall have more to say about this when we consider the Violaceous Trogons' nesting habits.

The song or call of the Violaceous Trogon is a long series of soft clear notes, all quite similar in tone and volume, uttered rapidly at a uniform rate or at most slightly accelerated toward the end. The notes of the Black-throated Trogon are rather similar, but weaker, more hesitant, and typically only two to four in a series, so that the songs of these two yellow-bellied species are readily distinguished. The songs of both of these trogons differ greatly from those of certain other sympatric species, as described in the preceding account. Violaceous Trogons are most vocal in the dry season when several males may gather to sing close together in the treetops, evidently competing for mates. In the Valley of El General, where the drier weather begins in December or January, I have heard their mellow *cow cow cow cow* . . . chiefly during the first quarter of the year but Violaceous Trogons may sing sporadically until late September when rains are heavy. While they prepare their nests the partner who rests nearby while the other is at work continues to utter similar notes in a soft, melodious undertone.

Very different from the song is the complaint or alarm call. One day I watched a number of birds protesting the presence of a Spectacled Owl that rested somnolently high in a forest tree, its feathers all puffed out, making it appear huge in comparison with the complaining birds around it. Among them was a male Violaceous Trogon who repeated sharp, rattling notes, with each utterance slowly raising his tail well above his back. This gesture contrasted with that of a male White-tailed Trogon who fanned out his tail feathers sideways, with a very rapid movement, each time he voiced his less strident complaints. The trogons did not approach nearly as close to the big owl as several hummingbirds did. A similar dry rattle is given by Violaceous Trogons disturbed at their nests.

Nest

In southern Central America Violaceous Trogons begin to nest in the dry season early in the year. In the Valley of El General at an altitude of 3,000 feet, a pair was excavating a nest cavity on 15 February. On Barro Colorado Island in the Canal Zone eggs have been found on 25 February (Eisenmann, 1952:28). In El General I have watched parents feed nestlings in early June. In the Canal Zone breeding evidently continues even longer, for Eisenmann noticed a pair building in June-July.

Violaceous Trogons are catholic in their choice of the material in which they excavate their nest chambers. Although other trogons such as the Massena and the White-tailed may nest in either a decaying trunk or a termitary, the Violaceous is the only trogon I know that adds a third medium, the papery material of a large wasps' nest. On Barro Colorado its nests have been found in a cavity near the top of a six-foot stump as well as in termitaries (Eisenmann, *loc. cit.*). Although in El General I once saw a male Violaceous Trogon show interest in a high decaying trunk, all the sustained nest preparation that I have seen, as well as all the occupied nests, were in vespiaries; a total of 10 were claimed by the trogons. These were all of the same type although apparently not all of the same species of wasp: very large turbinate structures composed of silvery gray paper, attached by the broad top to a more or less horizontal branch of a tree. Such nests have the entrance at the contracted lower end and contain numerous horizontal combs of hexagonal cells, one sheet above another, all composed of the same papery material as the envelope that covers them. Such vespiaries are usually built on a high branch; those in which I have seen Violaceous Trogons nesting were from 15 to 100 feet up, with the majority between 30 and 50 feet. It was apparently in a similar nest that Wetmore (1968:417) noticed that a "pair had made an opening in the upper end of a huge wasp's nest 20 meters from the ground in an open-limbed tree," near Almirante on the Caribbean coast of Panama near the Costa Rican border. The vespiaries occupied by Violaceous Trogons were either at the forest's edge or, more often, at a distance from woodland, in a pasture, coffee planation, cane brake, or else in a streamside tree.

As in other trogons the nest chamber is excavated by both sexes but the male often takes the initiative. On 15 February 1936 I discovered a pair of Violaceous Trogons digging into a wasps' nest of the type already described, hanging near the end of a branch of a small guava tree that grew in a thicket in the valley of the Río Buena Vista in El General. This vespiary, at least a foot long, was no longer occupied by its makers. The papery envelope was broken in several places, exposing the brood combs within, and it looked old; but perhaps the trogons were responsible for its recent abandonment and ragged aspect, since they may damage a wasps' nest while they decimate its occupants, as will become evident in the following pages. The birds had begun to dig into its sloping side at a point near the lower end, as they usually do. Clinging alternately in the upwardly directed hole they had made, back outward and downward, they tore at the brood combs with their bills. While one was at work the other rested in the guava tree, constantly repeating a low soft note, a subdued version of the *cow* call. Their spells of work were short, from less than one to five minutes. After I had watched them for 40 minutes the male flew away and his mate followed. This ragged vespiary apparently did not suit them, because as far as I saw, they never resumed work on it.

FIG. 9. Vespiary in which Violaceous Trogons raised three young. Rivas, Valley of El General, Costa Rica, May 1936.

Probably it was this same pair that 12 days later attacked an occupied vespiary of the same kind attached, a hundred feet above the ground, to an outer limb of the wide-spreading crown of a noble *Stryphnodendron excelsum* tree growing at the forest's edge. In the early morning the trogons flew out of the forest to perch amid the fresh new foliage of the great tree. They had made, or found, a tiny hole in the side of the vespiary that faced outward from the tree and their efforts were directed to enlarging this orifice. This was a difficult undertaking for, with no place to cling, they were obliged to work while hovering upon beating wings, which they could continue only for a second or two. Suddenly leaving its perch on an outer twig, a trogon would dart up to the vespiary, bite the edge of the aperture once or twice, rarely thrice, then alight on a convenient branch. This method of working, with wings rapidly beating and yellow belly showing brightly, reminded me of the trogon's manner of plucking a berry from a tree. Standing at least fifty yards from the nest, I sometimes heard the sound of the bird's bill striking against its hard surface. Darting up to it alternately, the two partners made dozens of attacks on the vespiary but they did not accomplish much. When, after more than half an hour of strenuous effort, they flew back into the forest, the gap that they were trying to make was not noticeably larger than when they arrived. I judged it to be about an inch in diameter.

On the following morning the trogons arrived at 06:45, worked for 42 minutes, making as little progress as before, then returned to the forest. On the third morning they worked less, probably because they were disturbed by a third trogon calling just within the woodland's edge. On the fourth morning they started to work at about 06:30. The female discovered that she could cling to the lower edge of the slowly widening gap in the vespiary's side while she tore at the upper edge. Clinging with her tail pressed against the bottom of the vespiary, she bit and tugged at the tough silvery gray paper of the envelope with movements so vigorous that her whole body shook. Crackling sounds reached me far below whenever she succeeded in breaking off a piece of the material. Now she could tear at the vespiary for several minutes together. Slow in following his mate's example, the male continued for a while to nibble ineffectually at the edge of the aperture while hovering momentarily on wing. Soon, however, he began to cling and bite the tough paper just as she did. Now they were taking fairly equal turns at the task. While one worked the other rested nearby, repeating soft notes in an undertone. They tore at the vespiary for a full hour before they returned to the forest.

While the trogons worked early in the morning, wasps were so little in evidence that, for a while, I thought the vespiary was abandoned. Apparently, however, they were merely torpid and inactive in the cool early morning at 3,000 feet above sea level; as I have seen at close range in the wasp *Synoeca surinama* on chilly February mornings even a thousand feet lower: many fell to the ground when they tried to fly. By the middle of the forenoon, when the bright dry-season sun had warmed the air, the black medium-sized wasps high in the *Stryphnodendron* tree were active enough, crawling over the outside of their home, flying in and out of their doorway, and gathering in numbers about the aperture that the trogons were making, probably to repair it.

After their spell of work on the fourth morning that I watched them, the trogons went off through the forest, but at about nine o'clock they returned to the vespiary. Now, instead of enlarging their hole they attacked the active wasps. Perching farther from the vespiary than they had done while watching each other work in the cool early morning, they made long spectacular darts to catch the insects, sometimes seizing them in the air, sometimes plucking them from the surface of their home. A sharp tap rang out each time a trogon's bill struck the vespiary in picking off a wasp. Although I clearly saw the birds catch the insects, I could not see just what was done with them. Even with eight-power binoculars I could never detect anything in a trogon's bill when it alighted on a perch after making a capture. Probably the wasp was swallowed in the air, as I never saw one fall. The female held her bill open much of the time, as though to cool her mouth after swallowing something hot, but possibly she was only panting after her strenuous exertion. For half an hour the pair of birds continued this meteoric mode of catching wasps, the while uttering pleasant low *cow*'s. Then, after an interval in the forest, they returned to catch more wasps but they did not resume excavation while I watched until late in the morning. Possibly now the very active wasps would have punished them.

Even while the trogons were catching the wasps late in the morning, the latter rarely attacked their persecutors. Once, however, I distinctly saw a wasp pursue a trogon as it flew from the nest after picking off one of the inmates. Later I watched the male trogon pluck from his abdomen a fluffy feather to which a wasp had attached itself. The wasp then flew away but the feather stuck to the bird's bill. He tried to scratch it off, then to rub it off against his perch, but it stuck persistently. Finally, he solved the difficulty by swallowing the feather.

One might suppose that the wasps would have defended their home with greater zeal, but they seem not to attack their natural enemies with the same fury that some kinds direct against blundering humans and domestic animals. When I watched a Red-throated Caracara tear open a corrugated nest of big black Guitarrón wasps (*Synoeca surinama*), armed with barbed lances that are reputed to inflict very painful stings, the wasps hardly made an effort to protect their home and the pupae and larvae that the caracara was devouring. And wintering Summer Tanagers make a practice of eating the immature stages of smaller wasps, apparently with impunity.

After working for six or seven mornings the trogons had opened in the side of the vespiary a gap into which their body fitted. Now, when they clung upright in the side of the structure while they continued to dig into it, only their long tail projected beyond the gap. They were now tearing into the brood combs, working alternately in shifts that continued for from two to five minutes. Although it was difficult to see just what they did with their head inside the vespiary, certain movements that I glimpsed suggested that they were swallowing something. I had no doubt that they ate the tender white wasp larvae and pupae which caracaras, Summer Tanagers, Scarlet-rumped Caciques, and other birds eagerly seek. Thus the vespiary not only provides the trogons with a shelter for their eggs and young, it supplies them with much food while they prepare to nest. Even at this stage when the trogons were digging into the heart of the vespiary, some adult wasps remained.

Neither trogon would work in the absence of its mate. The one who arrived first would perch near the vespiary and continue to call until its mate arrived, then one would go to the structure and proceed to dig into it. Similarly, if one flew off into the forest when the other wished to continue with their task, the latter would perch near the vespiary, doing nothing but call—once the female did so for 21 minutes after her mate left. In refusing to work at the nest unless the mate were nearby, trogons agree with motmots, kingfishers, and jacamars but differ strikingly from woodpeckers, in which either sex will continue to carve the nest cavity while its mate is out of sight. I could not learn just when these trogons completed their chamber in the vespiary. By 27 March, a month after I found them attacking the wasps' nest, they had not begun to incubate.

I was led to the discovery of the first vespiary into which the trogons were digging by following the voice of a Piratic Flycatcher that, in mid-February, had just arrived from its winter home farther south. While the trogons hollowed out the occupied vespiary high in the great *Stryphnodendron* tree, a pair of Piratic Flycatchers spent much time in this tree, calling with breezy insolence. Sometimes one would dart at a trogon, but at this stage they worried the larger birds little, as though instinctively aware that to do so now would only defeat their own ends. After the workers had finished their task, however, the pirates became more menacing, often darting over the trogons' heads as they approached their nest. Mostly the trogons took little account of their tormentors, hardly even ducking their heads as the smaller birds swooped by; but occasionally a trogon would lose patience and dart toward the pestiferous flycatchers, causing their immediate and hurried retreat. Presently the latter would return and continue to act as though they owned the nest. Finally they did gain full and undisputed possession of it, doubtless by the ruse of throwing out the trogons' eggs as I have seen them do at nests of other kinds of birds (Skutch, 1960a). The flycatchers nested in the vespiary but apparently without bringing forth a brood. Before they moved in the last wasps had vanished.

Three years later, on the lower side of the Valley of El General, I watched another pair of Violaceous Trogons attack an occupied wasps' nest. This vespiary hung below an upper branch of an *Inga* shade tree in a small coffee plantation about 35 feet above the ground. It had much the same shape as that in the great *Stryphnodendron* tree but the middle-sized, yellowish wasps were evidently specifically distinct from the darker insects that inhabited the other. This, together with the warmer dawns which occur later in the year and at a lower altitude, doubtless accounts for the somewhat different course of events at this vespiary. Instead of being quiescent until after sunrise, like the wasps a thousand feet higher in February and early March, these yellowish wasps were quite active before sunrise in late May and June.

Soon after daybreak on 27 May 1939 a pair of Violaceous Trogons arrived and started to catch these wasps. From perches of about equal height in neighboring trees the birds darted swiftly past the vespiary, sometimes skimming its surface, sometimes shooting past the entrance about which the wasps were swarming. Without much doubt they were catching and eating the wasps, but their movements were so swift that it was difficult to obtain visual evidence of this. Only once did I see a trogon swallow something after alighting on a branch; I did not see wasps fall from their bills, as would have happened if they merely tried to exterminate the inhabitants of the vespiary. Usually, after darting past the wasps' nest the trogons came to rest at a safe distance, but sometimes they did not go far enough and pursuing wasps would cause them to flee. Or else the trogon would try to beat off the attacking wasps with its wings. While resting between flights the birds constantly voiced low, soft notes.

For the next two weeks I found the trogons catching wasps every morning that I went to watch them. They would start in the dim early light, continue for about an hour, and leave about the time the sun rose above the distant mountaintops. Apparently, they made a satisfying breakfast on the unfortunate wasps. I first noticed the trogons attacking wasps later in the day on 9 June at about 14:30 in the afternoon. On the following day I found the birds catching wasps around noon. The trogons pursued the same tactics as in the early morning; but now, with higher temperatures, the insects were more active. Sallying forth in pursuit of their tormentors, they often caused a trogon that perched too near their home to retreat.

During the 16 days that elapsed between the date when I first noticed the trogons catching wasps and the start of excavation, the vespiary was slightly damaged. The covering of silvery gray paper about the doorway at the nether end had been badly frayed by the frequent light impacts of the trogons, exposing the lowest brood-combs. Higher on the sides of the envelope were some small rents, evidently made unintentionally by the birds as they plucked wasps from the surface of the vespiary. Accordingly, when one finds Violaceous Trogons nesting in a somewhat tattered vespiary it is not safe to assume that they had found it already deserted; they may have themselves damaged the structure. I have watched trogons feed nestlings in vespiaries that had lost most of their envelope, leaving the curved sheets of brood-cells largely exposed. A round entrance penetrating these combs led to the chamber where the young were concealed.

At last, on 12 June, I noticed that the trogons had made a shallow round depression in the side of the vespiary, the first step in excavating their brood chamber. On the following morning I spent three hours watching them at work. The two partners were digging into the structure at different points! The female had started her hole on the side facing outward from the supporting tree, where the entrance to an occupied chamber is usually situated. Her mate chose to work on a lateral face, parallel to the supporting branch. Although the two worked alternately in plain view of each other, each persisted with its own undertaking. Would the pair finally raise their family in a chamber with two doorways or would they ruin the vespiary?

The male's excavation was more advanced than that of his mate. On the morning of 13 June he already worked with head and shoulders inside while she still was on the outside. The reason for this became evident as I watched. He took longer spells, lasting from seven to 11 minutes, while hers were much shorter, not exceeding two minutes. Their labors were accompanied by the usual soft, melodious subsong. Doubtless because he worked harder, the male tired sooner and went off. His mate wished to continue but, as usual, would not work in his absence; after two more very brief visits to the vespiary she followed him into the distance.

By the following morning, 14 June, the male was working with his whole body inside the vespiary and only his long tail projecting; the female still clung to the side with most of her body exposed. When the male's turn came, he worked continuously but the female broke her turn into several short spells, as is evident from the following record of their activities from 07:00 to 08:20 (intervals in minutes, unless otherwise noted):

Female: 2, 1, 1, 1, <1, <1, 1, 1 = about 8.
Male: 11, continuously.
Female: 3, 1, few sec., few sec., 1, few sec., few sec., few sec. = about 6.

Male: 21. He might have worked longer but the female, growing impatient, alighted on the nest and made him fly out.
Female: 3, 2, <1, 1, <1, 1 $=$ about 8.
They fly away.

A few wasps lingered in and around the vespiary while the trogons were at work, but they seemed dispirited and rarely bothered the birds. The female, nearer their doorway and more exposed, was more subject than her mate to their infrequent attacks. Sometimes when they hovered around her head and shoulders she turned back her head and snapped at them. Rarely one attached itself to her plumage, making her dart rapidly away trying to shake or pluck it off, which she usually succeeded in doing promptly. If a wasp ever stung a trogon, I saw no evidence of it in the form of obvious distress or swellings. Rarely now a trogon snatched a wasp from the air. Once I clearly saw one member of the pair crush a wasp in its bill and drop it to the ground. On the second morning that I watched the trogons dig into the vespiary the wasps were still fewer and did not molest the birds at all.

While this pair of trogons was catching wasps at the vespiary into which they later dug, neighboring pairs of Violaceous Trogons were already feeding nestlings. Whether because of the lateness of the season or because they ruined the vespiary by making holes on two sides, they had not started to incubate when I left them at the end of June and I doubted whether they would do so. This pair caught wasps early in the morning and, after the number of these insects had been reduced, dug into the vespiary late in the forenoon; the first pair did just the reverse. Without additional studies it is impossible to say which procedure is more usual.

Eggs and Young

The only nest low enough to be reached with a high step-ladder already held young when found, and I have not seen the eggs of this trogon. Belcher and Smooker (1936) mention two addled eggs collected on the Guanoco River in Venezuela. They were dull white and measured 31.3 by 23.3 and 29.7 by 23.5 mm. An egg in the British Museum from Cachí, Costa Rica, is likewise dull white with a faint gloss and measures 29.2 by 22.6 mm. (Wetmore, 1968). The set may consist of three eggs, as I have seen this many nestlings.

I have no observations on incubation in this species. Since the male broods the nestlings, he doubtless incubates also, probably sitting most of the daytime, as in other trogons. The female broods and, therefore, undoubtedly incubates by night.

Five vespiaries in the Valley of El General, 2,000 to 3,000 feet above sea level, held nestlings between late April and early June. The nest with young to which I gave most attention was found on 22 April

1936 in a vespiary attached to a small tree standing above a brake of the giant cane *Gynerium sagittatum*. It was about 30 feet up in the treetop and quite inaccessible, but by standing below and peering through binoculars up into the wide downwardly directed doorway, I could dimly discern some of the lighter parts of its occupants: the bill and yellow orbital ring of the male; the female's bill and the white spots before and behind her eyes; and, later, the bills and, very vaguely, the heads of three nestlings. In contrast to trogons that nest in open niches in low trunks and are easily alarmed, these Violaceous Trogons felt secure in their high, enclosed abode. The brooding parents remained within while I moved noisily beneath them over crackling dead canes; even shaking the slender tree and pounding on its trunk failed to make the female emerge. Both parents fed their nestlings while I watched below them, unconcealed.

When the male arrived soon after daybreak on 23 April, the female left the nest, where she had brooded through the night, with her tail badly rumpled from long sitting in the confined space. For several minutes he clung in the doorway, only his rump and tail showing, while he fed nestlings that were evidently still very small. Then he climbed inside to brood them. While the male trogon was in the dark chamber a black Chisel-billed Cacique hopped across the top of the vespiary, causing him to dart out. Perching nearby he repeated, over and over, a low rattle, slowly elevating his tail with each repetition of the complaint. Finally, he changed to the clear *cow cow cow* . . . and flew away. The cacique looked on, then dived into the thicket.

The nestlings received both insects and small fruits. After they were older they sat with their heads above the orifice, uttering almost constantly a low, sweet-toned *cow cow cow cow,* much like that of their parents but more subdued. As each repeated these notes its throat swelled out, separating the growing feathers and revealing the skin beneath them as a lighter point amid the darkness. Sometimes they leaned so far over the opening that their yellow under plumage was visible from below. From time to time they regurgitated seeds that fell through the doorway. They remained in the nest for 17 days after I found them. On 9 May I discovered a newly departed fledgling resting on a low perch amid the tall canes and repeating the same low melodious *cow cow cow* that I had earlier heard from the nest. After permitting a very close approach, it flew so weakly that I easily overtook and captured it. In plumage it rather closely resembled its mother. The two outer tail feathers on each side were white, at least as far as exposed, and the remaining rectrices were black. Its bill was bluish gray with a lighter tip; the bare legs blackish and the toes dark flesh-color. Each dark brown eye was surrounded by a ring of dull yellow bare skin, behind which was a crescent of whitish feathers, and in front a smaller crescent of the same color, much as in the adult female.

After I had examined the fledgling and set it on a perch, its mother came to feed it close to me.

Summary

The Violaceous Trogon inhabits forest edge, thinned woodland, and open country with scattered tall trees, from lowlands up to about 4,500 feet. In heavy forest it remains in the treetops. Its food consists of both insects and small fruits snatched while hovering on wing.

Its song is a long rapidly delivered series of similar soft notes, all of the same pitch. It complains with a dry rattle uttered while slowly elevating its tail.

In southern Central America the Violaceous Trogon nests from February until June or July. In Panama it has been found nesting in cavities in decaying trunks and in termitaries, but in the Valley of El General, Costa Rica, all 10 nests studied were in large turbinate wasps' nests, covered with silver-gray paper and attached high in trees.

Occupied vespiaries are often, if not always, chosen by the trogons. Their procedure differs with the species of wasp and the climate. At 3,000 feet above sea level in February, when nights were chilly, a pair of trogons started to break into the vespiary in the early mornings while the wasps were sluggish. Later in the morning when rising temperature had activated the wasps, the trogons caught them on the wing and apparently ate them. A thousand feet lower, in May and June, when with warmer nights wasps of a different kind were active at daybreak, a pair of trogons caught them chiefly before sunrise. After this had continued for more than two weeks and the population of the vespiary had been reduced, the trogons started to break into it.

The two sexes work alternately at carving a brood chamber in the heart of the vespiary, but the male usually works harder. At one vespiary the male and female started to dig in different places. Neither will continue to work unless the other is nearby. Apparently they devour whatever larvae and pupae they find in the brood combs. Remaining wasps sometimes attack the trogons, but without much spirit; the birds seem to suffer little inconvenience from them.

The only nest of which the contents could be ascertained contained three nestlings, who were fed and brooded by both parents with the female occupying the nest by night. The young remained in the high vespiary for at least 17 days and on emerging rather closely resembled the adult female.

Family ALCEDINIDAE

RINGED KINGFISHER

Ceryle torquata

About 15 inches long, the handsome Ringed Kingfisher is by far the largest and most powerful of the six species of this family in the Western Hemisphere. In the male the upper parts are slate-blue, the wings and tail largely of the same color, barred with white. He wears a conspicuous crest and a broad white collar around his neck. His breast and abdomen are rich cinnamon-rufous. The female differs chiefly in having her chest slate-blue, separated by a narrow white band from the cinnamon-rufous of the more posterior under parts. In both sexes, the long, strong bill is blackish, the eyes are brown, and the legs and toes are dark.

The Ringed Kingfisher has an enormous range from northern Mexico to Tierra del Fuego, from the Pacific Ocean to the Atlantic over most of this vast extent, and in the Lesser Antilles. It occurs casually in southern Texas. Unlike many birds with so wide a geographical range, it has a restricted altitudinal range. I have from time to time seen one of these birds flying high above the rushing mountain torrent in front of our house at 2,400 feet, apparently en route to or from a deeper pool upstream. Among the few published records of its occurrence at higher altitudes is that of Wetmore (1968:422), who found this kingfisher at 4,200 feet in western Panama. It is understandable that this big kingfisher, which needs fairly deep water for its dives, should be most abundant in the lowlands where such bodies of water are chiefly found. But scattered through tropical America at middle and even high altitudes are deep lakes and broad streams which seem to offer ample scope for its fishing. Perhaps falling temperature sets an upper limit to the Ringed Kingfisher's vertical extension; yet a species that has successfully colonized such an inclement region as Tierra del Fuego should be able to adapt itself to the mild climates of middle altitudes within the tropics, as the far smaller Green Kingfisher has done. Perhaps man's persecution of this large and conspicuous bird in the altitudinal belt where human population is densest may be the reason why we so seldom find it there.

These kingfishers live chiefly along deep, smoothly flowing lowland streams and the shores of lakes and lagoons. They may frequent brackish or even salt water, as in the innumerable canals and tranquil arms of the sea along the island-studded coast of southern Chile (Goodall, *et al.*, 1957:318). They avoid the narrow, deeply shaded forest streams where the smaller kingfishers are sometimes seen. Watercourses are the natural highways of these kingfishers, which they are frequently reluctant to leave. One day, ascending in a motor launch the Toloa Creek in northern Honduras, we drove a Ringed Kingfisher before us for possibly a mile. Each time that the boat approached he would leave his perch, fly a few hundred feet ahead, and alight on another branch overhanging the stream. Here he would wait until the launch was almost beneath him, then fly ahead a few hundred feet more. Only after this procedure had been repeated many times did he finally double back over the bank and return downstream. The Amazon Kingfishers, the Little Blue Herons, and to a less extent the Middle American Jacanas along this stream had the same habit, which is widespread among waterfowl. On the other hand, Ringed Kingfishers sometimes have overland routes between widely separated bodies of water and traverse them, especially in the early morning and in the evening, flying high with regular wing-beats, delivering at intervals of a few seconds the single stentorian *kleck* which accompanies their flight.

Ringed Kingfishers scratch their head by raising a short leg inside, or over, the drooped wing.

Food

Ringed Kingfishers, like the other American members of the family, subsist chiefly upon fish, often of fair size, which they catch in the usual manner of their tribe. Being larger birds their plunges into the water from an overhanging branch are more spectacular than those of other species. After a successful dive they carry their victim to a branch, against which they knock it vigorously before they swallow it headfirst. Sometimes a kingfisher delays a great while to consume what it has caught. Once I saw a female alight in a balsa tree beside a stream with a fish half as long as herself dangling crosswise in her great bill. For more than 2½ hours, by my watch, she held the fish, shifting her position only from one branch to another of the same tree. After this delay she started to beat the fish against a branch, although it must long since have died.

Voice

The Ringed Kingfisher has a very limited vocabulary, consisting of hardly more than a single call, a hard *kleck*, which is modified in various ways to meet the various occasions of its life. A single loud *kleck*, repeated at measured intervals, is the almost invariable accompaniment of its flight. A softer, rapidly repeated *kleck* is the signal that a kingfisher is ready to replace its mate on the eggs. A very loud, rapid, mechanical rattle or series of *kleck*'s expresses anxiety or anger, as when the young appear to be in danger.

Nest Building

It is advantageous for kingfishers to nest in the drier part of the year, when streams are lowest and least likely to rise and flood their burrows in the banks or wash them away. In the Caribbean lowlands of northern Central America Ringed Kingfishers begin to dig their burrows in February, if not earlier. Along the Río Morjá, a tributary of the Río Motagua in northeastern Guatemala, I found a nearly completed burrow on 19 February 1932 and on 22 February another which was apparently finished. Still another burrow, discovered in March, was evidently started at about the same time as the first, for the eggs in both hatched on the same date. These three burrows were in low, vertical banks of sandy loam which had been freshly cut during the preceding rainy season when the river ate into the banana plantation along its left shore. The entrance of one tunnel was 20 inches below the top of a bank which rose six feet above the water at its low stage. The others were similarly situated.

Beginning long before their eggs will be laid, the kingfishers proceed in a leisurely way to dig their tunnels in the soft alluvial loam and one may have to wait many hours to see them at work. Early in the morning of 21 February I concealed myself across the river from the first burrow, hoping to learn something of the details of excavation; but it was 11:00 before I heard a measured *kleck, kleck, kleck* and, turning, saw a kingfisher approaching from upstream. It soon entered the tunnel and remained within for a few minutes, appearing not to notice the Rough-winged Swallow who fluttered in front and several times rested in the entrance while the kingfisher was busy inside. Emerging, the kingfisher perched atop a banana leaf and kept up a running conversation of low rattles with its mate, out of sight around the bend upriver. In five minutes it flew upstream to join the other, calling loudly *kleck, kleck, kleck.*

Soon the pair returned together and began to work in earnest. Each time that one entered the burrow a jet of earth was thrown outward from the mouth. As the bird moved inward the jet followed until it passed from view. Doubtless the bird continued to kick the earth backward until it reached the head of the excavation, and thus the material loosened by its bill was gradually pushed out of the tunnel. The kingfishers invariably emerged headfirst, indicating that the burrow had reached its final length and had begun to widen at the far end into the nesting chamber where they had room to turn around. The bird inside called with low *kleck*'s, which were answered by its mate who perched on a rusty tram rail that had been washed out by the flood and now leaned against the riverbank just below the burrow. The mates seemed to encourage each other in their dark subterranean labor. Both sexes shared equally in the toil and as soon as one emerged and flew up beside the other on the rail the latter entered the burrow, kicking out jets of earth as it disappeared into the darkness. Four or five minutes was the usual interval spent in the earthwork. Once both were together in the burrow for a few minutes. Just after noon, while one kingfisher was engaged inside, its mate seemed to tire of waiting on the rail and flew upstream. When the other emerged and found itself alone it followed. Although I continued to watch until the middle of the afternoon, the kingfishers did not return, having worked less than a hour that day.

When completed this burrow was seven feet three inches long. It curved slightly in both the horizontal and vertical planes so that, looking into the front with a flashlight, it was impossible to see what was at the inner end. The entrance tunnel was six inches in horizontal diameter by four inches high. The top of the enlarged chamber at the inner end was 23 inches below the surface of the ground. Another burrow, about 400 feet downstream, was seven feet 8½ inches long and its chamber was 22 inches below ground. The third burrow, higher upstream, was shorter, only six feet seven inches long. When opened, while incubation was in progress, the first two burrows had no lining; the eggs rested on the bare, sandy floor of the brood chamber.

Eggs

Each of the two burrows that I opened contained four eggs which must have been laid early in March, for they hatched at the end of the month, after an incubation period that was probably not less than that of the Amazon Kingfisher, which is about 22 days. The ovate eggs were pure white and glossy, similar to those of woodpeckers. The measurements of eight eggs averaged 45.4 by 32.6 mm. Those showing the four extremes measured *46.8* by 32.1, 46.0 by *34.1*, and *43.3* by *31.0* mm.

According to Bent (1940:135) the Ringed Kingfisher "apparently lays three to six eggs." In Chile, where this kingfisher nests in November and December, the set is said to consist of three to five eggs (Goodall, *et al.*, 1957:318).

Incubation

Much as I would have liked to have opened the burrows before the eggs were laid, in order to learn the length of the incubation period, experience with other burrow-nesters had taught me that they are apt to desert if one meddles with their nests before incubation is well begun. Accordingly, I waited several weeks before I prepared the two most conveniently situated burrows for observation of what was happening within them.

My studies of Amazon Kingfishers and Turquoise-browed Motmots had shown that the best way to prepare a burrow is to dig down behind it and make an opening in the back of the brood chamber just big enough to permit one to reach the eggs. After each inspection the opening should be closed with a stone or a board and the pit one has dug carefully covered over and concealed. The first step in my undertaking was to measure the length of the burrow by inserting a flexible woody vine. This snake-like intrusion failed to drive out the incubating kingfisher. But when, after measuring back the indicated distance from the top of the bank, I began to dig almost above him, he flew out, uttered a few *kleck*'s, and headed upstream, where he perched on a giant cane leaning over the water and soon plunged for a fish. I now saw that the bird I had driven from the nest was the male. Like other burrow-nesting birds in the same circumstances, he seemed rather unconcerned about what I was doing, although later I found him to be the most devoted of parents.

As I broke into the brood chamber the buzzing of green flies, making a peculiar hollow resonance in the long burrow, told me that incubation had begun; for the kingfishers, sitting on their eggs, regurgitate the indigestible bones and scales of the fish they eat, and these attract the carrion flies. Lying flat on the ground, I could barely reach the four white eggs that rested on the bare, sandy floor of the brood chamber, more than two feet below the surface. Even before I closed the aperture that I had made at the back, the male kingfisher started to reenter the burrow. When he noticed the unaccustomed light at the rear he fled with considerably more agitation than he had shown when first disturbed.

To learn the incubation pattern of these kingfishers was one of the most challenging and perplexing ornithological problems that I ever tried to solve. At that time little accurate information was available on incubation patterns anywhere, and my own experience in studying them was quite limited. Yet I had already seen enough to be convinced that incubating birds do not sit capriciously but follow, even if loosely, some sort of schedule. Despite certain perplexing observations at the outset of my study of these kingfishers, I was certain that they must fall into a pattern, if only I could discover it. The great difficulty was to keep my attention so firmly fixed on a hole where for long hours nothing happened—a sort of yogic exercise in the contemplation of nothingness—that I did not miss the sudden, unannounced exit of a kingfisher. Then, too, it was often difficult to distinguish the sex of the bird who entered or left, for it might go in or out without allowing me a fair view of its breast, where alone there are conspicuous differences in the coloration of the male and female. To ease the burden of long vigils when no kingfisher was in sight, I set a little twig in the mouth of the tunnel. So long as it remained upright, I was assured that no bird had passed the doorway.

I soon discovered that there was a single changeover each day, in the early morning. This was comparatively easy to observe, for the kingfisher coming to take its place in the burrow usually flew downstream sounding its powerful, metallic *kleck, kleck* at measured intervals, and so heralded its own arrival. Then it perched on a banana leaf overhanging the bank, where it was often difficult to pick out, for the gray of its plumage blended with the gray of the undersides of the leaves; but the low, rapid rattle that it now uttered helped to reveal its presence. It might swoop down toward the hole in the bank, as though about to enter, but continue its flight unbroken and alight on the midrib of another frond on the other side of the burrow and there continue to call. These false starts were often repeated three or four times before the bird finally gathered courage to go in. About a minute and a half or two minutes elapsed before this bird's partner emerged from the burrow, gained the middle of the river, paused an instant as it turned its course upstream, then flew low above the channel, klecking, until it had passed from view beyond the willows at the bend in the stream. The critical point in my observations was the instant when the emerging kingfisher plunged toward me as I sat on the opposite shore of the river. If I failed then to notice the color of its breast, I might remain ignorant of its sex, for a fraction of a second later it had turned its unrevealing tail toward me.

After I had overcome the difficulty of learning the sex of the kingfisher leaving the burrow, I found that on some mornings it was the male who entered and the female who left, while on other mornings the reverse was true. There was a regular alternation, the male entering on one morning and the female on the next morning, as the record shows:

March 17, 07:51. The male replaces the female.
18, 08:29. The female replaces the male.
. . . .
21, 07:33. The female replaces the male.
22, 10:01. The male replaces the female.
. . . .
25, 09:01. The male replaces the female.
26, 08:04. The female replaces the male.
28, 07:05. The female replaces the male.

Two irregularities in this record require explanation. Since the male replaced the female on 17 March, we should expect him to do so on all odd dates, but on 21 March we find that the female relieved the male. This shift was without much doubt the result of the nest's remaining unattended through the night of 18-19 March, when I had disturbed the female in order to make sure of her sex. It was evidently she who returned to the nest on the following morning. The break between the 22nd and the 25th was also, I believe, caused by my interference with the nest when I was puzzled by an observation.

Each afternoon the incubating partner took a single recess, for food or exercise, from its long 24-hour turn on the eggs. It emerged suddenly and without warning, at some time between 13:00 and 16:00, flew upstream to the feeding ground, leaving the nest unoccupied, and returned in from half an hour to an hour. On returning it flew downstream low above the water and entered the nest directly, without perching or calling, in a manner quite different from the morning entry, since there was now no mate in the nest to be advised of its arrival. Then it remained on the eggs continuously until relieved by its partner on the following morning.

One day the female took her afternoon outing early, or possibly omitted it, for there was no movement into or out of the burrow between 14:55 on 21 March and 10:01 the following morning, at which hour the male, arriving late, relieved her. Thus she had incubated continuously for no less than 19 hours or, if she had omitted her afternoon excursion, for the 26½ hours that had elapsed since her entry at 07:33 on 21 March. Sixteen or 17 hours was the usual interval between the kingfisher's return in the afternoon and its release by its mate on the following morning.

In the great majority of inland birds the rhythm of incubation follows a 24-hour cycle so that, in its main features, one day is much like the next, not just the reverse, as in the Ringed Kingfisher. This shorter cycle is followed by the smaller kingfishers, including the Amazon (Skutch, 1957b), the Green, and the Half-collared Kingfisher of Africa (Moreau, 1944). Although rare among inland birds, a 48-hour cycle, such as is exemplified by the Ringed Kingfisher, is not unusual in seabirds, among which it is followed by the Diving Petrel and several terns. Indeed, many marine birds, including numerous species of penguins, albatrosses, shearwaters, and other Procellariiformes incubate continuously, while fasting, for days or even weeks together (Skutch, 1957a, and references there cited).

The kingfishers displayed a surprising degree of formalism in their behavior. One morning I arrived at the river in the gray dawn and after visiting the burrow to make sure that the sentinel stick had not been pushed over since the preceding afternoon, waded the cold water of the Río Morjá. I took my station in the grass clump on the opposite shore. At 07:30 the male shot out of the burrow without warning and flew upstream as usual. I was surprised by this unconventional behavior, but probably no more than his mate. Five minutes later she appeared from the direction of his departure and perched in a cecropia tree growing on the riverbank a few yards from the burrow, where she often rested before going to replace the male on the nest. Soon he returned and the female, who had not moved, greeted him with a rapid, low rattle, which I interpreted as a scolding. A minute later he flew upstream again. I now fully expected the female to enter, since it was her day to take charge of the eggs; but such an unwelcomed entry would evidently have been a breach of formality. She delayed another minute, as though pondering what course to follow, then flew off after the delinquent. Soon both returned, but separately, and there followed a rather lengthy dialogue between the pair as they rested on banana leaves near the burrow's mouth. Finally, after flying back and forth several times before it, the male entered while his mate continued to *kleck* in a low voice from her perch.

The female kingfisher had won the dispute. She delayed a seemly interval for him to settle on the eggs, then after giving the customary warning, entered herself. The male came out in record time (about one minute after his mate's entry) and turned upstream, klecking loudly, a free bird at last.

Both partners were extremely attached to their nest and as incubation proceeded it became increasingly difficult to drive them from the burrow, which I rarely attempted to do, for I wished neither to disturb their routine nor to risk their desertion. When I removed the stone that closed the aperture at the rear of the burrow to see whether the eggs were about to hatch, the bird on duty merely retreated into the tunnel, where it stayed until the hole was again closed.

Five days before the eggs hatched one member of the pair gave me a supreme example of devotion. Having failed during the customary hours to witness the morning changeover, I was eager to learn which partner was with the eggs and thought that perhaps I could drive it far enough into the tunnel to see its breast from the front and so recognize its sex. I removed the stone from the rear of the chamber, then hurriedly jumped down the bank and looked into the entrance, but no kingfisher was visible. Climbing up the bank again I lifted out the eggs and listened for the tapping of the chicks within, but still the parent did not retreat far enough into the tunnel to be seen from the front. Next I directed the beam of the flashlight into the brood chamber from the rear. Five days earlier this had sent the female rushing out in a panic after no amount of tapping on the stone at the rear, or even removing it, would make her retreat. But the bird now on duty was not easily alarmed. Finally, all other expedients failing, I thrust the lighted flashlight itself into the chamber, into which, of course, I could not see, for the pit I had dug was too narrow to admit my shoulders. The kingfisher dealt the shining object a resounding blow and retreated backward. Hurrying to the bank I found its quivering form blocking the tunnel so near the entrance that I reached in and gently touched its finely white-barred tail, which was turned outward. I might have pulled out the bird by its legs but I dared not take that liberty; yet without seeing its breast I could not tell whether the male or the female was before me. Fifteen minutes later the heavily breathing body still remained where I had left it. It was the female's day on the nest but I shall never be sure to which partner I should give credit for such steadfastness.

Meanwhile, I had found another Ringed Kingfishers' burrow, about 400 feet downstream in the same bank. I opened it only two days before the eggs hatched. The bird attending them, with a degree of attachment that I have rarely seen equalled, remained bravely in the burrow while I probed its length with a vine, dug a pit in the rear, removed the eggs for measurement, fitted a stone in the aperture that I had made, and tamped the soil above it. All this took well over an hour, during which the other partner did not appear. The one on duty remained firmly in the tunnel, where from the front I could see its great, black beak half agape, its crested gray head with a white spot in front of each eye, and its white collar, but not enough of its breast to distinguish its sex.

The territories of these two pairs of kingfishers extended in opposite directions from their burrows. The attendants of the upper nest always approached it from upstream and returned thither when relieved, while those of the lower nest fished downstream from it.

Nestlings

On the morning of 28 March the female of the first nest arrived in its vicinity at sunrise, an unusually early hour for her appearance. She had not been in the burrow for 24 hours and could hardly have known whether the two eggs that I had found pipped on the preceding evening had hatched, but she was somehow excited. She flew around in circles and had some inconsequential clashes with her neighbors downstream, although up to that time I had never seen the two pairs come into contact. At 07:05 she entered the burrow—the earliest changeover that I recorded. After waiting long enough for her to settle in the burrow, I opened it at the rear and found that two of the eggs had hatched; the other two were pipped.

These eggs were first fractured by the enclosed chicks in several distinct places lying in a single octant of the surface, between the greatest transverse diameter and the thick end. I saw no intermediate stages but when the nestlings finally emerged the cap, which they pushed off the shell, was asymmetric, the line of separation being strongly oblique rather than transverse as in many passerines. The newly hatched kingfishers had pink, peculiarly transparent skin, without the least vestige of down. Their eyes were represented by two relatively huge protuberances that projected above the forehead. In the center of each the sightless pupil was dimly visible through the skin. The lower mandible projected about $\frac{1}{16}$ inch beyond the upper, which was blackish in its central portion. The skin on the neck and chin was peculiarly wrinkled. In profile these newly hatched kingfishers were certainly not handsome! On each heel-joint was a thickened callous pad, neither smooth as in jacamars, motmots, and nunbirds (*Monasa*), nor strongly spiked as in woodpeckers, toucans, barbets, and other birds that nest in holes in trees or termitaries, but covered with small papillae. Resting on their entire feet with the protected heel joints, with additional support from their abdomens, these kingfishers less than 15 hours old could already stand erect and even walk a little in a tottering fashion. They made a low, sizzling sound.

Replacing the nestlings in their burrow, I walked down the riverbank to look into the neighboring nest. When I removed the stone from the rear a sizzling noise arose from the earth, for three of the eggs had hatched. Two of the chicks retreated with their mother into the tunnel, beyond my reach. In both nests the eggs had hatched on the same day.

The males of the two pairs, both of whom happened to be free that day, behaved differently than on previous mornings. After being relieved by their mates they had gone off as usual to their respective territories up and down the river; but instead of remaining there, as they had always done before the eggs hatched, they soon returned to perch near their nests and loudly protest my presence. In their excitement, both perched at the same time in the cecropia tree that grew on the riverbank between the two nests and was apparently on the boundary between their territories. They stood side by side on a branch, their beautiful, white-barred, slate-blue wings spread until they almost touched. One raised his crest but the other laid his flat, and with open bills and angry *kleck*'s each defied the other to cross the accepted frontier.

On this morning I heard an utterance from the kingfishers that I had never noticed before. It was really not so much a new note as a different way of using the old familiar one, for their whole vocabulary apparently consists of a sound which to my ear is suggested by the syllable *kleck*, which is variously modified in diverse situations. Now that they had nestlings to guard, the parent kingfishers expressed their anxiety by a very loud, rapid, mechanical klecking, continued, with momentary pauses, as long as danger seemed to threaten. This harsh, deafening rattle was uttered while the bird perched with its mandibles parted and motionless, its tail vibrating rapidly up and down. The loud rattling was even continued in flight, when the usual flight note, the stentorian *kleck*, seemed to be superimposed upon it, with an odd effect. This din was hardly diminished by the presence of a fish in the parent's bill. I never heard another bird make more noise when its nestlings seemed to be in danger nor give more evident signs of distress. Yet, in common with other birds that nest underground, the parent kingfishers never darted at me nor made any other demonstration. They merely perched in full view and rattled interminably.

I hardly required any further demonstration of the parental devotion of these kingfishers, but some destructive meddler provided an opportunity for them to show the full strength of their attachment. When I approached the first burrow two days after the eggs hatched, the parents were perching nearby, the female with a fish in her bill, both rattling as fast as they could. Hurrying to the burrow, I found that my entrance at the rear had been enlarged until it occupied the entire end of the brood chamber. The nestlings had evidently retreated into the tunnel so that the rogue could not reach them from this end, for he went around to the front and cut away at the entrance with his machete until it gaped like the jaws of an alligator. After all, he did not want, or could not reach, the nestlings, for all four remained; but one was pale, cold, and unable to stand. I closed off the back of the burrow as well as I could and replaced the nestlings. Three eventually succumbed but the parents continued to feed and brood the fourth in the exposed and gaping burrow.

Stuffed with whole minnows to the bursting point, the single survivor grew rapidly. When it was a week old the rudiments of its body feathers were visible as dark points beneath its skin, from which, three days later, they were emerging. The eyes of the ten-day-old nestling were almost fully open and it uttered a high-pitched trilling sound in response to its parents' rattle. When it was 14 days old the upper mandible equalled the lower in length and the contour feathers began to project from the tips of their long sheaths. The young kingfisher now tried to defend itself and bit hard with its great mandibles whenever I picked it up. A few days later it squealed and fought like a fury. When about 24 days old it was fully feathered, except the naked belly that rested upon the sandy floor of the foul

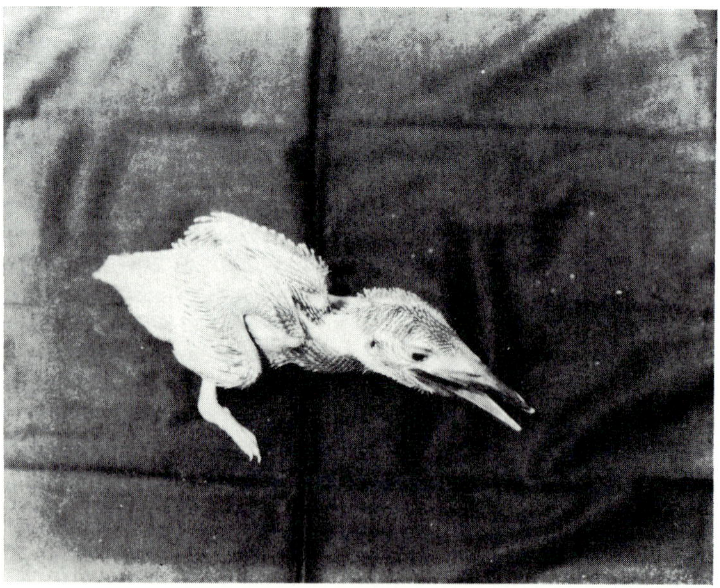

FIG. 10. Nestling Ringed Kingfisher, two weeks old. Near Los Amates, Motagua Valley, Guatemala, 10 April 1932.

burrow. When four weeks old it could at last perch on a stick but it still did not attempt to flutter, and when placed on the ground it could do no more than flap along with outstretched wings. To its biting and squealing the nestling now added an alarm rattle almost as loud as that of the parents, which finally drew one of them, who answered in kind. This young kingfisher remained in the burrow about ten days after it was fully feathered, finally leaving when between 33 and 35 days old.

Although this nestling was one of the most vociferous that I ever knew, its struggles and cries were mild compared to those of the single young kingfisher who survived in the burrow downstream. Looking into this burrow about two weeks after the eggs hatched, I was unable to reach the nestling from the rear or to glimpse it from the front so I concluded that it had fallen prey to some marauder. Two more weeks passed before I again visited this burrow to learn whether Rough-winged Swallows had occupied it; I was amazed to behold the young kingfisher, now almost as big as its parents, staring at me from the far end. On my preceding visit it had evidently escaped my notice by remaining just at the bend of the tunnel. Now, when I tried to reach it from the back it fled through the tunnel and jumped into the river, where it spread its wings, turned upstream, and flapped its way slowly against the current. When it encountered obstacles in the form of stranded brush, it hooked its bill over them and scrambled across. Thus it led me a lively chase, wading in the muddy shallows, until a fallen banana plant halted its wayward progress and I seized it. Unlike its neighbor of the same age upstream, it had not become accustomed to being taken in hand; its deafening screams and fierce attempts to bite made, for duration and intensity, the best efforts of the other pale to insignificance. Ten agonized minutes passed before it became reconciled to me and perched on my hand until its feathers dried. Then I replaced it on the filthy floor of its burrow, swarming with white maggots and exhaling a strong odor of ammonia. Of all the nestlings that I have ever handled only the Red-billed Tropicbirds were equally pugnacious and noisy.

This fledgling left its burrow when between 35 and 38 days old. The Ringed Kingfishers remained in their nests about a week longer than the smaller Amazon Kingfishers and ten days longer than the still smaller Green Kingfishers. In plumage, both birds whose development I followed resembled the adult females, but the slaty feathers of their chests were conspicuously edged with chestnut. Between the slate-blue of the chest and the chestnut of the posterior under parts one had a white band, which the other lacked.

Early in the morning of 15 May a young Ringed Kingfisher came to the mouth of the burrow which I had not opened. Pausing there with its head and breast alone visible, it uttered at intervals a loud

FIG. 11. Nestling Ringed Kingfisher, four weeks old. Near Los Amates, Motagua Valley, Guatemala, 26 April 1932.

rattle. As it bobbed its head up and down in typical kingfisher fashion, the nearly level rays of the sun, just rising above the foliage at the bend upstream, fell upon its broad collar, which gleamed with snowy whiteness. At length the fledgling launched forth, turned downstream, and rose to perch in a willow tree at the mouth of a small tributary. On its very first flight it covered about 200 feet and rose 30 into the air.

SUMMARY

Over a vast range, Ringed Kingfishers live chiefly along the deeper bodies of fresh, brackish, or salt water, at low altitudes. Rarely they ascend as high as 4,000 feet. They subsist largely upon fish.

They have a limited vocabulary, consisting of a loud, hard *kleck*, modified in various ways.

In Central America these kingfishers nest early in the year, when rivers are low. Dug by both sexes, working alternately, their burrows in loamy river banks may be eight feet long. The enlarged chamber at the inner end is not lined. In the Caribbean lowlands of Guatemala two sets, each consisting of four pure white eggs, were laid in early March. In other regions sets of from three to six eggs have been recorded.

The male and female incubate for alternating periods of about 24 hours. The single daily changeover usually occurs between 07:00 and 10:00, when the partner arriving to begin incubation enters the burrow before its mate leaves. Each afternoon the kingfisher interrupts its long spell of duty by an outing that lasts from about 30 minutes to an hour, during which the eggs remain unattended.

The newly hatched kingfisher has no trace of down or feather rudiments on its transparent pink skin. Its lower mandible projects about 1/16 inch beyond the upper mandible. Each heel is covered by a callose pad, studded with low papillae, which protects it from abrasion as the young bird stands on the unlined floor of the nest chamber.

At the age of 10 days the nestling's eyes are open; at 14 days the upper mandible is equal to the lower in length and the contour feathers begin to project from their long sheaths; at about 24 days the young kingfisher is well feathered; at from 33 to 38 days it flies from the nest with power and control. When handled, older nestlings defend themselves furiously.

The parents fail to clean the nest, which soon swarms with maggots and exhales a strong odor of ammonia.

Ringed Kingfishers are most devoted parents, remaining steadfastly with their nestlings in the face of much disturbance; yet, in common with other burrow-nesters, they neither threaten intruders nor attempt to lure them away by distraction displays.

Family BUCCONIDAE

WHITE-FRONTED NUNBIRD

Monasa morphoeus

The most memorable event of my visit to the Río Yavarí in 1940 was my first meeting with nunbirds. On the shallow-draft gunboat *Amazonas* of the Peruvian inland navy, our rubber survey party had voyaged over the Amazon and its great tributaries in transandean Peru from the Pongo de Manseriche in the west to the Yavarí on the frontier of Brazil. On the left bank of this southern tributary of the Amazon we went ashore to search for rubber trees at a tiny settlement which someone with a longing for cool breezes and a sense of humor had named Islandia (Iceland). While we examined a small rubber tree growing in a plantation near the forest, a little flock of blackish, starling-sized birds perched not far above our heads, twitching their tails from side to side while they uttered soft, musical murmurs. They had large, dark eyes and bright orange bills that contrasted with their dusky plumage. Although they were quite unlike any bird I had ever seen, their zygodactylous feet, with two toes directed forward and two backward, assured me that they were not passerines. Later I learned that they were Black-fronted Nunbirds, belonging to the puffbird family. My duties as botanist of the survey party did not permit further observation of these strange birds, but this brief encounter convinced me that they had unusual social habits that would well repay careful study.

Although for the next quarter-century my bird watching was done in parts of Central and South America where nunbirds do not occur, my desire to study them, born of this transitory meeting, never quite died away. My next sight of nunbirds was in March of 1965 on a short visit to La Selva, in the Sarapiquí lowlands of northern Costa Rica, where the White-fronted Nunbird is found. I was able to pass two months at La Selva during the breeding season of 1967. Although we saw much of the nunbirds, the only nest that we then found was prematurely destroyed by a predator. Returning in March of the following year we struck better weather and better luck, finding three occupied nests, two of which were successful. They amply confirmed my surmise, held for so long, that the sociable nunbirds have extraordinary habits.

Nunbirds, of which four species in the genus *Monasa* are currently recognized, are named for the sombre, nun-like attire which all wear. The Costa Rican race of the White-fronted Nunbird is a fairly stout puffbird about 11 inches long. In both sexes the upper parts of the body and the wing coverts are slate-colored, the under parts lighter slaty gray. The remiges and the rounded tail, of moderate length, are nearly black. The black head is adorned with short, stiff, outstanding white feathers on the forehead, lores, and chin. Projecting from this facial ruff is a bright orange-red bill, of moderate length, that tapers from a fairly broad base to a sharp tip. (The glossy, orange or reddish bill of certain South American nunbirds has earned for them the name *pico de lacre*—lacquered bill.) The large, soft eyes are brown. The short tarsus and toes are blackish. As in a number of other puffbirds, females average larger than males in most of the conventional measurements (Ridgway, 1914), but the difference in size of the sexes is too slight to permit their separation in the field.

The White-fronted Nunbird ranges from eastern Honduras to Peru, Brazil, and Bolivia. A heat-loving bird, in Costa Rica it is confined to the Caribbean lowlands and foothills and appears rarely to ascend above 1,500 feet above sea level. Here as at La Selva, it inhabits wet forests whose dominant trees are immensely tall and heavily burdened with a vast variety of epiphytic growths. Although they usually remain well up in the trees, nunbirds on occasion descend to the ground for food, as they do to nest. More often than amid uninterrupted forest, nunbirds are noticed around the edges of, and within, openings which facilitate insect-catching. Often they venture a good way into clearings and plantations with isolated tall trees, such as cacao plantations with scattered shade.

Nunbirds travel through the treetops and across clearings in small parties of rarely more than half a dozen individuals, which are evidently family groups. These birds do not fly in a compact flock, like certain parrots, but straggle along at intervals, in the manner of toucans and certain cotingas. Often they associate loosely with other more or less social birds of the treetops. At La Selva we often watched them foraging among the shade trees of the cacao plantations with Purple-throated Fruitcrows and Scarlet-rumped Caciques, whose plumage is also predominantly dusky, and such other birds as Rufous Mourners, Cinnamon Woodpeckers, Rufous-winged Woodpeckers, Black-striped Woodcreepers, and Massena Trogons. Late on May afternoons we often found a certain cacao plantation, adjoining the forest, alive with such mixed aggregations of actively foraging birds. We never saw a nunbird behave aggressively toward any other bird, of its own or another species.

The flight of nunbirds is direct and strong but rarely long continued. Sometimes, when crossing an open space, they set their spread wings and glide. Although like other puffbirds they perch motionless while looking for food, their more social nature prompts greater activity. They tend to remain for shorter intervals in one spot, rarely puff out their plumage conspicuously, and do not give the impression of inert dullness which is responsible for a whole array of scientific and popular names denoting stupidity and sloth which have, unfortunately, been applied to this remarkable family.

Food

Puffbirds of all kinds rarely eat vegetable food, and the nunbirds are no exception to this rule. In many hours of watching, both while they foraged and while they fed their young, we never saw one with a berry or other fruit. They subsist almost wholly on invertebrates, with occasionally a small vertebrate, such as a tiny frog or a small lizard. The great bulk of their food consists of insects, including many cicadas and a variety of orthoptera, among which grasshoppers, mantids, and phasmids are prominent. Sometimes a beetle or dragonfly is taken. Many of the insects are in the larval stage. Among other invertebrates that are consumed are spiders and millipedes.

Nunbirds forage in the energy-conserving manner typical of puffbirds: they remain on some convenient branch, scanning their surroundings until their keen eyes detect some suitable prey flying in the air, creeping over the bark of a tree, or hiding amid the foliage; they make a sudden swift sally to seize their victim and carry it to a perch before they devour it. Sometimes when they dart out from a treetop to seize flying insects their mode of foraging resembles that of the related Swallow-wing. In clearings where well-spaced trees offer convenient lookouts, they catch many small creatures amid low herbage or even from the ground, ascending to a low perch to swallow each victim. The drying grass and weeds in a freshly chopped plantation or dooryard are a source of readily-caught insects and spiders of which nunbirds are quick to take advantage. Often we watched them foraging in the narrow clearing, between the forest and the river, in which the house at La Selva stood. Their sharp eyes could detect a green insect amid green grass at a distance of 60 feet or so. After carrying their prey to a perch in their bills, they would often beat or rub it against the branch; but their pounding was rarely as vigorous as that to which certain flycatchers, motmots, jacamars, and other birds subject large prey, and usually they gulped down their insects with the wings still attached.

Late one morning in April I watched three nunbirds foraging with army ants (*Eciton*), near a burrow in which they were feeding nestlings. The nunbirds perched from about four to 15 feet above a broad forest pathway over which the ants were swarming. At intervals they alighted on the ground to capture some insect that the ants had stirred up, or else they plucked it from a tree trunk, or even overtook a flying fugitive. After each sally they promptly returned to a perch to watch. Repeatedly they looked at their feet, as though to make sure that no ants were crawling over them. Some of their prey was carried off toward the nest.

Voice

Some years ago, when the only puffbirds that I knew well were a few Central American species, I wrote that "Puff-birds are among the most silent of the birds of the tropical forests of America." (Skutch, 1958b:212). Since becoming familiar with the Double-banded Puffbird in Venezuela and White-fronted Nunbirds in Costa Rica, I regret having generalized too hastily. Some puffbirds are quite vociferous and the White-fronted Nunbirds have such a varied repertoire of calls that for a while I was inclined to attribute to them any loud, unfamiliar bird note that reached my ears at La Selva. Usually, if I could trace the sound to its source, my surmise proved to be correct. Among the vocalizations of which the significance finally seemed clear are the following:

1. The flight note, a low, soft monosyllable, uttered one or more times as a nunbird takes wing.

2. The call-for-companions, a series of loud, full, "mournful" notes, delivered by a lone perching bird separated from its companions, as by one who has just emerged from a long spell in the nest and does not know where its flock-mates are. One morning in April I watched a solitary bird resting on an exposed dead twig at the very top of a tall living tree. It called *how how how how* . . . in a soft, melancholy voice, finally rising to a crescendo *how how how how how*.

3. The approach call, so-named because it was most often heard from birds approaching their burrow with food for nestlings. This call has many variations but it seems always to consist of an emphasized opening note followed by a series of more rapidly uttered lower notes, producing an undulatory effect, which on different occasions I described as a soft roll, a liquid ripple of sound, a slow trill, a soft purr, a churr, a rattle between wooden and liquid. As the nunbird begins this call, it first elevates, then depresses its tail, trogon-like. A bird hesitating to descend to its burrow with food, because it is being watched, may repeat this call over and over for many minutes together. If the food-bringer is perturbed, the trill may become lower, softer, more prolonged—a melodious wail of remonstrance. But the call is by no means confined to occasions when feeding is delayed by a human observer or some other potential danger; one may hear it coming from the vicinity of a nest while he stands a good way off, unseen by the nunbirds. Indeed, a nunbird preparing to take food to nestlings may give this call sparingly while still hundreds of yards from the burrow and again when it breaks its journey to rest in a tree midway of its flight. The call is heard less often from adults feeding young already on the wing. Occasionally it is given by a solitary nunbird not bearing food; then it seems to be used as a substitute for the call-for-companions. At times this call resembles that of the Barred Antshrike.

4. The feeding call, a rapid series of low, sharp notes uttered by a

nunbird standing at the mouth of a burrow with food for the nestlings in its bill. Often the food is delivered with no sound audible to an observer at a distance of 40 or 50 feet; but if the young delay to come for their meal, the food-bringer begins to call, the notes growing louder the longer it must wait, until they become a sharp *click click click*. . .

5. The begging or hunger call. As the nestlings grow older they may utter, as they come to take their meal from an adult at the burrow's mouth, a call that sounds like *tuwee tuwee*. This is especially likely to be heard when they have not been fed for some time and are very hungry. Fledglings waiting to be fed may perch on a branch and repeat these notes, in a high thin voice, for a long while. Adults soliciting food from their mate or other companions give calls which bear more or less resemblance to this juvenile hunger call, and seem to be derived from it. Often the adult's hunger call is uttered in a soft, whining or pleading voice, but at times it becomes piercingly shrill, especially as the hungry bird flies toward a companion to take the food from its bill.

6. The chorus is the loudest and most surprising of the nunbirds' many utterances. It is given by from three to ten birds perching close together, but not in contact, in the forest or sometimes in an adjoining clearing. The choristers may be stationed anywhere from about 25 to 75 feet up, occasionally higher, on a slender horizontal branch or, very often, on a vine stretched horizontally between two trees in the midst of the forest. Frequently they rest in a row, at intervals of a few inches and all facing the same way; but if numerous they may form two rows on neighboring branches or vines, and sometimes one or two perch singly, slightly apart from the others. Even when in a row, all do not invariably face in the same direction. One of these performances may last, with short intermissions, for 15 or 20 minutes, during which the birds from time to time change their perches and their grouping.

Perhaps "chorus" is too grandiose a term to apply to these performances, which in reality are no more than bouts of simultaneous shouting, with little rhythm and no melody. Tilting their heads upward, the dusky participants call all together in loud, ringing, almost soprano voices, with such vehemence that their whole bodies shake, making as much noise as a flock of chachalacas. Some relax their wings, or spread and close their tails, as they pour forth their high-pitched notes through their bright orange-red bills. At times one nods its head emphatically downward, as though in approval, or to encourage its companions to proceed with the "music." The chorus swells louder or wanes, as more or fewer birds join in, but for many minutes there are only the briefest intervals of silence. The shouting nunbirds appear to enjoy the chorus immensely; but if they are susceptible to sore throats, I imagine that they are so-afflicted when, after perhaps a quarter hour of almost continuous vocal effort, they fall silent and fly off through the forest in small groups or singly.

We heard these choruses through most of our sojourn at La Selva, from late March until early June, that is, from the outset of the nunbirds' nesting season until most, if not all, of their young had fledged. Probably they are by no means confined to this season. The choruses were especially frequent from mid-April until late May, when the young were growing rapidly in the nest or were already on the wing. We heard them most often late in the morning and around midday, sometimes in the early afternoon.

The nunbirds' choruses are not quite like anything else that I have heard in the avian world. In a way, they remind one of the social singing of certain gregarious birds, such as siskins, goldfinches, and Evening Grosbeaks, numbers of which sing simultaneously as they rest in a tree. But there is no coordination in the singing of such birds, each of which pours forth his medley as the spirit moves him, whereas the nunbirds' chorus is obviously a coordinated performance. In this it resembles the duetting or antiphonal singing of many constantly mated tropical birds, especially wrens, although it lacks the exquisite melody and the precise synchronization of many of these duets. Nevertheless, its function may well be the same, to strengthen the bond between the members of a family; only in this case the family consists not merely of a pair but of the three, four, or possibly more individuals who cooperate in caring for a nest.

This interpretation is supported by an observation that I made at the end of April. Hearing a chorus near the house, I entered the forest to seek the choristers. The sound moved from place to place, and presently five adults alighted almost above me on a slender, horizontal vine, 25 or 30 feet up. One held a large insect in its bill. Here they raised their voices again, and at the conclusion of this bout of calling, one, who had perched somewhat apart from the others, flew westward alone. Then the bird who held the insect flew eastward, followed by the three others in single file—an impressive sight. They alighted in a tree where these four nunbirds habitually paused before delivering food to the well-grown nestlings in a nearby burrow.

Two days later, in the early afternoon, there was another long-continued chorus, just within the edge of the forest near this same nest. Seven or eight nunbirds were present, and they joined in various shifting groupings, sometimes three, four, or five perching and calling close together, sometimes a single bird resting somewhat apart and remaining silent. I noticed no antagonism, but one alighted rudely almost on top of another, making it move. One nunbird called for a while with an insect in its bill but finally swallowed its food and continued. After about a quarter of an hour, when the excitement had almost died away, one bird gave the whining hunger call and another, who had been resting nearby, tried to mount her; but she flew away.

On 23 April I watched a chorus with eight participants, who at first perched close together on slender horizontal boughs, about 25 feet up in the forest. After a while they separated into two groups of four, lined up on branches about 20 feet apart, and continued to shout as before. After about 15 minutes the assembly dispersed, the birds flying off in different directions, leaving only two who remained to preen. This choral gathering occurred between a nest with young attended by four grown birds and another nest, where also at least four birds were in attendance, which had been destroyed by some predator about ten days earlier. Evidently these two groups or families composed the assembly. Possibly it had territorial relevance, although it was much nearer the nest with young than the nest which had been plundered. However, I have found no evidence of territoriality in nunbirds—nor, indeed, in any other bird that roams through the upper levels of tropical forest.

These are a few examples of the nunbirds' choruses that we witnessed. They vary considerably in circumstances while conforming to the same general pattern: the simultaneous shouting of birds perching close together, typically in a row with approximately equal intervals of a few inches between them. Although their biological function is not wholly clear, one carries away the impression that nunbirds indulge in them so frequently because they enjoy shouting together.

In Bolivia, Niethammer (1953:299) found four or more Black-fronted Nunbirds singing in chorus. He believed that they were sometimes stimulated to perform by hearing the shouting bouts of the Speckled Chachalaca.

Sunning

Late on a morning in May, when the sunshine was pouring down hotly into the clearing behind the house, three adult nunbirds flew down to lie on the lawn, where the grass had grown rather long. The one that I could see best had both wings widely extended. The birds snuggled down in the grass, preening and scratching in the brilliant sunshine. They were not shy and permitted me to watch them a few yards away. Soon they flew up into the forest trees that bordered the clearing.

Nunbirds scratch their heads by raising a foot outside a closed wing rather than over and inside a drooped wing, as most passerine birds do. The nunbirds' "direct" mode of scratching appears to be general among piciform birds. I did not see them bathe.

Feeding of Adults

When, after a morning in the forest, I returned to the house toward noon on 17 March, I was told that nunbirds had for some time been feeding young in the clearing. However, I could detect no sign of immaturity in any of the four individuals who composed the group. Late on the following morning the nunbirds again foraged among the scattered trees in front of the house. One of them repeated the hunger call much of the time, and this bird was fed thrice in 20 minutes by two others. Probably three members of the party were supplying food to the fourth but I could not keep them separate long enough to prove this. Doubtless one of the feeders was the recipient's mate, engaging in "nuptial feeding." But since it is improbable that the begging nunbird, evidently a female, was mated to two or three others, the term seems inapplicable to the whole activity. Sometimes the feeders brought the insect to her; at other times, she flew up to a companion who had returned to a perch after making a capture and, without alighting, snatched the object from its bill—just as dependent juveniles do. She also caught many insects for herself, quite expertly. Many of the small creatures devoured by this party were caught amid low herbage, to which the birds descended from low boughs or even a projecting root of an overturned tree. With these nunbirds were a Rufous Mourner and a pair of Great Kiskadees who perched quite near them and from time to time flew down to catch insects or spiders from the grass, exactly as the nunbirds did.

At the time when we witnessed this feeding, a female who nested in the forest near the house was evidently laying, or preparing to do so. Through the rest of March and the first week of April, we continued from time to time to hear the whining food call from one member of the flock that frequented the clearing, but it was fed more seldom than at first. After this the eggs hatched and the attendants, busy supplying the nestlings, rarely solicited food from each other and were never seen to receive any.

In late April and early May begging calls again sounded with some frequency among the trees near the house. Not only did the family nesting nearby forage here, but likewise another group whose nest was about 800 feet distant. This nest was destroyed in mid-April and I surmised that these renewed calls for food by an adult signified that she would lay again. But no replacement nest was ever found.

Nest

Three days after our arrival at La Selva on 10 April 1967, Dr. Dennis Paulson told me that he had found a nunbirds' nest. While studying dragonflies in a small grassy swamp he had seen three nunbirds fly across the open space from the neighboring cacao plantation, with food in their bills. Following them into the forest beyond the swamp, he soon found their burrow in a moderate slope beneath tall trees. Since the afternoon was already far advanced, we decided to visit the nest on the following morning. The next day dawned with rain and we waited until it abated before we set forth. To our dismay, as we approached the site of the burrow a heap of freshly dug earth greeted our vision. Some animal had dug out the first nest of the White-fronted Nunbird ever, to my knowledge, to be found by a naturalist, before even it could be properly described. Later I searched for a replacement nest on the surrounding slopes, but in vain. We found no other nest in 1967, although by late April fledglings were being fed in our study area.

In the following year we arrived earlier and began hunting nunbirds' nests in the second week of March, but the first was not discovered until 1 April, when it already contained naked nestlings. Another nest, also with nestlings at most a few days old, was found on 9 April, and the third, with older nestlings, was discovered on 21 April. These three burrows were revealed to us by the calls and movements of the attendant birds, which make the discovery of a nest relatively easy to the initiated, if he comes within hearing of the approach calls of nunbirds feeding nestlings. Of no other forest bird did we locate as many as three occupied breeding nests during our two seasons at La Selva. But it is most difficult to detect the narrow hole in the leaf-strewn forest floor, amid tangled undergrowth, without guidance from the nunbirds themselves. If while incubating they enter and leave the burrow only twice a day, as White-whiskered Soft-wings do (Skutch, 1958b), they are not likely to reveal its location during this period. Unfortunately, we found no nest until after the eggs had hatched. The burrow which we chiefly studied was only about 50 feet from a path along which we passed dozens of times while incubation was in progress. Indeed, I had searched repeatedly for a nunbirds' nest on the opposite side of the path, where the slope of the land was greater, as the only nests that I had hitherto seen were in such slopes.

The four burrows (including that prematurely destroyed in 1967) were in heavy forest, but only a short distance from a clearing or plantation in which nunbirds gather much of their food. They were in moderate slopes (up to about 30 degrees) or nearly level ground, and they descended at a slight angle from their mouth in the leaf-strewn forest floor. The first of the burrows found in 1968 was in a 30-degree slope on which many small palms, with cane-like clustered stems, grew beneath tall trees. It was situated below one of the clusters, and a large mossy log lay close in front of it. About 40 inches long and almost straight, the burrow descended into the ground with an inclination of about 20 degrees. Near its mouth the tunnel was three inches high and 3¾ inches wide, and it varied little in diameter until it expanded into the brood chamber at its inner and lower end. The bottom of this chamber was lined with many pieces of dead leaf.

The second burrow opened only 10 feet from the edge of a high, wooded bank beside the Río Puerto Viejo. Inland from the bluff the land inclined slightly downward, but the burrow entered the ground at a point where a small irregularity of the surface increased the slope. The mouth faced inland, away from the river. This burrow, which had only a slight downward inclination, was 55 inches long and nearly straight; but the enlargement at its inner end was offset to one side,

FIG. 12. Forest near a burrow of White-fronted Nunbirds. La Selva, Caribbean lowlands of northern Costa Rica, May 1968.

making it impossible to see, by looking in at the front while the tunnel was illuminated by a small electric bulb, what it contained.

The third burrow was in gently and uniformly sloping ground (less than ten degrees) in a little clear space amid the tangled undergrowth of heavy forest. It inclined slightly downward from the mouth and was at least 55 inches long, as I learned by probing with a slender stick. The outer part of the entrance tube was nearly straight, but toward the end it curved enough to cut off the view of the chamber where the nestlings rested. Near the mouth the tunnel was three inches in horizontal diameter, which varied little in the visible portion.

Since the tubular tunnel intersected the surface of the ground at a very acute angle, the opening of this third burrow was greatly elongated. An untidy collar of mostly coarse, decaying sticks, mixed with petioles and dead leaves, surrounded and reduced the oblique aperture, making it less conspicuous. Sticks that were evidently placed there by the nunbirds were up to 22 inches long, and some of the shorter ones were ½ inch thick; but the larger sticks were decayed and very light when dry. This collar was broadest and thickest above the elongated mouth of the tunnel, which it covered for a distance of five inches inward from the definitive opening, with a roof about two inches thick. On the sides and in front, the pile of sticks was thinner. The other two burrows, which intersected the surface of the ground at greater angles and accordingly had less elongated mouths, had thinner, less obvious collars, or scarcely any. At none of these burrows could the accumulation of sticks and trash compare with that over a burrow of the Black Nunbird found in Venezuela by Cherrie (1916:322). This burrow, 1.5 m. long, descended from heavily wooded, level ground at an angle of about 45 degrees from the horizontal. Over its mouth had been heaped half a bushel of rotten, coarse, dead twigs. From the edge of this pile a rounded tunnel ran along the ground to the mouth of the burrow proper.

In a number of points the nest of the White-fronted Nunbird resembles that of the White-whiskered Soft-wing, which inhabits the same forests but is much more widely distributed in Central America. Both are burrows in nearly level or moderately sloping ground, but those of the nunbird are two or three times as long as those of the soft-wing. Both of these puffbirds surround the aperture of their tunnels with a collar of twigs and petioles, and both line the brood chamber with dead leaves. Another similarity is that in neither case is any freshly excavated earth evident in front of the burrows at the time they are used for breeding, and the same was true of the burrow of the Black Nunbird found by Cherrie. We can only surmise why this is so. Do

FIG. 13. Entrance to White-fronted Nunbirds' burrow in the forest floor, showing the collar of dry sticks, petioles, and dead leaves around the opening in the ground. Near Puerto Viejo de Sarapiquí, Costa Rica, May 1968.

these puffbirds carry to a distance the earth they remove from their tunnels, as Prong-billed Barbets and certain other birds carry away the chips from the holes they carve in trees, and Yellow-eyed Juncos carry away in their bills the much smaller amount of earth they scratch out of the depression in the ground in which they build their nests (Austin, *in* Bent *et al.*, 1968:1129)? Do they, like certain motmots, dig their burrows half a year or more before they will lay in them, so that the earth thrown out in front, washed away or impacted by rain, mixed or covered with vegetable debris, is no longer evident when nesting begins (Skutch, 1945, 1964-c)? Do they not dig burrows for themselves, but depend upon holes they find ready made, perhaps altering them just enough to meet their needs? In many years of residence among forests in which White-whiskered Soft-wings nest I have failed to solve the riddle of their burrows. In at least one instance the same burrow was occupied in successive years; and in the measure that such reuse is general among ground-nesting puffbirds, the chances of discovering the origin of the burrows are correspondingly reduced.

Another unanswered question is whether nunbirds ever nest in burrows in vertical banks, as kingfishers, motmots, jacamars, and many other birds often do. On 22 March I saw a nunbird fly up and hover in front of a hole in a high, nearly vertical earthen bank of a meandering stream that flowed between the forest and a cacao plantation. After looking into the hole the bird turned away without alighting at its mouth. It did this again and again, then flew out of sight. Soon this or another nunbird came and the performance was repeated. After another absence, these flights up to the entrance, without landing there, were resumed. At least two nunbirds, of a group of four that a little while before had chorused in the vicinity, were doing this. Between flights to the hole, they perched low and voiced soft, soprano notes. The mouth of this cavity in the bank was much wider than high, and far larger than it needed to be for a nunbird to pass through. If it had been made by birds, perhaps kingfishers or motmots, it had been considerably altered since they finished it. When examined a few days after the nunbirds showed such great interest in it, there was no indication, in the form of scratches on the bottom, that they had entered the cavity. Unfortunately, after this episode I lost much time seeking nunbirds' nests in banks, where none was ever found.

Nestlings

Doubtless the White-fronted Nunbirds' eggs are pure white, as are those of the Black Nunbird (Cherrie, 1916:322), and indeed all puffbirds (and all members of the order Piciformes), as far as we know. As to their number, the only indication I have is that provided by the short burrow discovered on 1 April, which contained three recently hatched nestlings.

Appearance and development.—When I first saw these nestlings, by looking into the burrow with a flashlight, they were a few days old and dark feather rudiments were visible beneath their pink skin. They were devoid of down. Their eyes were closed and when they gaped widely I saw that the interior of their mouths was flesh-colored. So long as they remained in the chamber at the inner end of their burrow these nestlings were inaccessible to me; but on one occasion I was able to take one in hand. Nine days after we found this burrow, when the nestlings were probably about 12 days old, I was looking into it when an attendant alighted close behind me and trilled softly. Immediately the young started to run up the inclined tunnel and I easily caught the one that reached the mouth first. Its dark gray contour feathers were expanding quite generally, except on the crown, but much bare skin was still exposed. The flight feathers on wings and tail were still ensheathed. The nestling's eyes could open about halfway, but it seemed mostly to keep them closed. Its bill was light horn-colored. On its heels were prominent callous pads, smooth as in jacamars and motmots, rather than tuberculate as in kingfishers, or spiked as in nestlings which grow up in holes in trees or termitaries. When I replaced the young nunbird in the mouth of its burrow it promptly turned around and ran down to rejoin its mates at the lower end.

This was the only time that a nestling was so incautious as to permit itself to be caught by me. Usually they did not come to the burrow's mouth unless an attendant was already standing there with food, and after taking it they promptly retreated into the depths of the tunnel. Looking past the parent, I sometimes enjoyed a fleeting glimpse of a nestling as it took its meal.

When I found the long burrow at the top of the bluff on 9 April, the nestlings, whom I could see only when they came to take food, were still small and pink, evidently not more than two or three days old. The last nestling left this burrow on the morning of 7 May, when it was at least 29 days old, and probably a day or two older. Thus the nestling period of the White-fronted Nunbird is about one half longer than that of the smaller White-whiskered Soft-wing, which is only 20 days.

Brooding.—When I first looked into the short burrow on 1 April my electric light disclosed a well-grown bird in dusky plumage. Its head, with the revealing orange-red bill, was hidden; since it had not flown when we approached rather noisily, and remained while I examined the nest, even touching it with the end of the measuring stick, I supposed that it was a well-feathered nestling. When next I peered into the nest, however, only three naked nestlings were present. The feathered bird that I had first seen was a brooding parent, steadfastly remaining at its post in the face of our intrusion. I was reminded of the Double-banded Puffbird who remained bravely covering her single fresh egg while Cherrie (1916:321) cut and hacked at the termitary in which the nest chamber had been excavated.

On 5 April when we watched this nunbirds' nest all morning, the naked nestlings were left alone from 06:40, when the parent who had brooded through the night left them, until 11:29 when an adult entered to brood.

On 11 April we watched this nest from dawn to dusk. The parent who was brooding at daybreak left the burrow at 06:15, and for the remainder of the day the partly feathered nestlings were alone. As night approached no adult entered the burrow to brood. As later events proved, this was a most fortunate omission.

After brooding ceased in the burrow at the top of the bluff, I noticed at dawn that leaves from the floor of the chamber had been raised to form a low wall or screen, reaching halfway to the roof, in front of the nestlings, such as I was able to examine more satisfactorily in the case of short burrows of the White-whiskered Soft-wing (Skutch, 1958b:222-224).

Feeding.—At dawn on 9 April I stood at the top of the riverside bluff watching a burrow of the Broad-billed Motmot. Presently a nunbird's approach call drew my attention to one of these birds perching in a neighboring tree with a small walking-stick insect dangling from its bill. Acting upon this hint, I soon found the nunbirds' burrow a few yards from where I stood. Meanwhile, two other nunbirds had alighted near the one with the insect. Suddenly a large dead branch directly above the three broke off and fell, but they darted away just in time to avoid being struck. They promptly returned to the same small tree, where they were joined by three more nunbirds, making six adults in all perching close together. Evidently some of these birds belonged to another family but I noticed no antagonism between them. While I examined and measured the nest, most of the birds flew off; but soon two returned with food, making three that waited with offerings for the naked nestlings. After I withdrew to watch from a distance, two flew down to feed the nestlings, but the third finally carried away its insect.

Many hours of watching at this burrow throughout the month that the young remained in it convinced us that four grown birds were feeding the nestlings. This was apparently the group of four that we had earlier seen giving insects to one of its members in the neighboring clearing. Frequently we saw the four together near the nest, and often three were carrying food simultaneously; but it was more difficult to assure ourselves that all four were attending the young. At the short burrow in the hillside four nunbirds were also bringing food. At the third nest found in 1968, with an unknown number of nestlings already well grown, at least three nunbirds were bringing food. It will be recalled that, in the preceding year, Paulson was led to the discovery of the ill-fated first nest by seeing three birds approach it with insects in their bills.

Aside from the tattered tails of some, which helped us to recognize them as individuals, there were no obvious differences between the three or four attendants of a nest. All appeared equally mature. One might suppose that these nunbirds were nesting communally like anis (*Crotophaga*), two pairs sharing a burrow. However, the only nest at which the number of eggs or young could be ascertained, without digging out and probably ruining the burrow, contained only three nestlings of about equal size and age, which seemed to be the offspring of one female rather than two. I believe that each burrow belonged to a single pair and the extra attendants were helpers, probably yearlings who would not themselves reproduce until a later nesting season, as in White-tipped Brown Jays and Banded-backed Wrens (Skutch, 1960a), apparently also Collared Araçaris (Skutch, 1958a), and certain other birds (Skutch, 1961b). To demonstrate beyond doubt the relationship of the several attendants of a nest would necessitate the banding of nestlings and following them through the succeeding year or more. Such a study would encounter great difficulties because it is so hard to find burrows and to reach the nestlings without breaking up the nests, and because the short legs of nunbirds are mostly hidden beneath them while they perch, which would make it difficult to see the bands.

While attending nestlings, nunbirds are far more tolerant of a human observer than are many other birds of lowland forests, especially those whose nests are low. We watched them while sitting or standing 30 or 40 feet from the burrow's mouth, with little or no concealment by the vegetation. At the short burrow at least some of the attendants would deliver food while we stood only four or five yards away. Some of the nunbirds hesitated longer than others to approach the nest in our presence; but this was not wholly a disadvantage, for while they delayed others might arrive with food, and such piling up of attendants helped us to learn how many there were.

We never saw a nunbird approach the nest with more than a single item of food, usually held in the tip of its vivid bill. Usually this was of substantial size, especially after the nestlings were some days old. Insects, which provided the bulk of the nestlings' diet, were nearly always delivered with their wings still attached; there was little or no beating or rubbing to detach them after the attendant arrived in sight of the burrow. Very frequently the nestlings received cicadas, which were also a major item in the diet of young Broad-billed Motmots in neighboring burrows. Orthoptera, some of which were green and quite large, were frequently brought. Other items that we recognized were beetles, spiders, caterpillars, rarely a small frog or lizard, only once a large and conspicuous butterfly or moth, once apparently a *Peripatus*. Once, to my amazement, an attendant gave to a half-grown nestling a large, green caterpillar covered with branched stinging spines, such as a man cannot touch without being punished with the most acute pain. I had doubts about the wholesomeness of such food, but when I looked into the burrow three hours after it had been swallowed, none of the nestlings appeared to be unwell. The Squirrel Cuckoo sometimes eats such stinging caterpillars and feeds them to its little ones.

Flying up with food for the nestlings, often from a neighboring clearing, the attendant nunbird never went directly to the burrow but always alighted in a nearby tree, usually from 15 to 30 feet up. Here, flagging its tail up and down, it gave the rippling or churring approach call, already described. As a rule, this call was repeated over and over for about two or three minutes; rarely it was given only once or twice or even omitted. Some of the more timid or distrustful attendants would continue to call for far longer, occasionally as much as half an hour, and perhaps finally fly away with the food undelivered. This delay, spent in calling, doubtless gives the nunbird an opportunity to scrutinize the surrounding forest and assure itself that none of its hereditary enemies (among which man is probably not included) is lurking in sight. Finally assured, the attendant flies down and alights at the burrow's mouth. Usually it delays a good fraction of a minute (possibly repeating the feeding call in a voice too low to be audible to us a dozen yards away), before a nestling comes to the entrance and takes the food from it. If the delay is prolonged the sharp notes of the feeding call are heard, and they become louder, more insistent, the longer the attendant must wait. A few seconds after delivering the meal the attendant rises into the trees, to fly away promptly, or to wait for the companions who have accompanied it to the nest to deliver what they have brought and fly off with it. On only one occasion did we see two attendants stand at the burrow's mouth with food at the same time, but one of them flew up again without having delivered it. Only when it intended to brood, which was seldom, did an adult enter the burrow with food.

Even the youngest nestlings that we found, still blind and naked, took their meals at the burrow's mouth, as do White-whiskered Softwings from a very early age. They ran up the tunnel bobbing their sightless heads and waving their rudimentary wings. In the case of the longer burrows of the nunbird, the round trip from the chamber to the mouth and back again was eight or nine feet, a long pedestrian journey for a tiny mite that crawled out from beneath a brooding parent and then pushed back into its sheltering warmth. Apparently the nestlings do not ordinarily respond to the approach call, which seems loud enough to reach them even at the end of a burrow four or five feet long, no matter how often this call is repeated. If they did they should be waiting at the burrow's mouth when the attendant descends to it, but nearly always the feeder must wait on the ground for a nestling, and sometimes it must entice it outward with a loud feeding call.

Nearly always a single nestling appears in the doorway to take the food, then backs out of sight as soon as it has seized the object. More rarely two nestlings come to the doorway, although only one can be fed. This is especially likely to happen with very young nestlings. Doubtless, as they grow older they learn that it is useless to go to the doorway behind a nest-mate. Older nestlings sometimes give the begging call, *tuwee, tuwee,* as they take their meals. Except when coming for food, nestlings of all ages remain discreetly out of sight in the depths of their long burrows, and they are rarely heard.

Exceptional behavior is exhibited by nestlings who appear to be very hungry. On 8 April rain fell almost continuously throughout the day and there was no gleam of sunshine. One of the nunbirds who foraged in front of the house perched with drooping wings and a badly frayed tail, begging for food in a whimpering voice—the very image of dejection. I feared that the nestlings would suffer, but on the following morning those in the short burrow on the hillside were very much alive. They sat looking outward, repeating soft little *peep*'s, while I peered in at them with a flashlight. Soon an attendant bearing a cicada alighted about four yards from the burrow and repeated the approach call. Hearing this, at least two nestlings moved up to the entrance and stood there, exposed and uttering sharp, insistent notes for a minute or so before the attendant flew down and fed one of them. After the adult flew away the nestlings delayed in the doorway for another minute. Even on 10 April, after a day of abundant feeding, these nestlings were noisier than those in deeper tunnels, and one ran into my hands at the burrow's mouth, as already told. The behavior of these nestlings, noisily exposing themselves at the surface of the ground, seemed most imprudent, and perhaps was the reason why disaster overtook them a few days later.

Although there was much daylight in the forest by 05:30, the attendants brought little food, and sometimes none at all, before 07:00, and on some days it was 08:00 before they fed actively. The nestlings received rather infrequent but substantial meals. On 2 April, when the three nestlings in the short burrow were about four days old, they were fed only four times in the three hours between 08:25 and 11:25, during much of which interval rain fell hard. On 5 April these same three nestlings received 12 meals in the six hours from 05:40 to 11:40. On this morning rain fell until about 9:00, after which the sky cleared. On 11 April, when these three nestlings were becoming feathered, they were fed 30 times in the course of the first rainless day that the month brought forth, from 05:40 to 18:00. Twenty of the meals were brought in the forenoon and 10 in the afternoon, only four of them after 14:00. On 10 April an unknown number of naked nestlings in the burrow beside the bluff were fed 11 times in the six hours from 05:40 to 11:40, on a morning with intermittent showers. Some of the attendants of this nest were more timid than any of those at the short burrow, and carried away food that they had brought for the nestlings.

Behavior of attendants who lost nestlings.—I returned to the short burrow late on the morning of 12 April to be greeted by a gaping hole with a pile of freshly dug earth in front of it. Some powerful animal, probably a Tayra (*Tayra barbara*) had plundered the nunbirds' nest. The three nestlings had vanished without a trace. Fortunately, for the first time they had passed the night alone, and no adult had been with them to share their fate. From the brood chamber I raked out a handful of fragments of dead leaves, amid which many small white larvae wriggled. Although attendant nunbirds do not remove droppings from their nest—indeed, seem never to enter it after brooding ceases—I noticed little waste matter. Probably the larvae had helped to disintegrate it.

While I examined the ruined burrow, four attendants arrived with food. One soon vanished but three perched in neighboring trees for a long while, holding their intended contributions and repeating soft approach calls. After I withdrew a short distance, they went, one by one, to stand in front of the gaping hole and call the vanished nestlings with sharp feeding notes. One did this while I stood only a dozen feet away. A melancholy spectacle! Finally, the nunbirds flew off toward the neighboring clearing, each holding the food that it could not deliver.

On the following morning, 13 April, no nunbird came near the pillaged nest while I watched from 08:30 to 09:30. As I was leaving I heard the approach call from a bird in the distance and hurried back to my observation post. The nunbird continued to call in the trees above me and after a while I managed to see it. It held a large cicada in its bill, while it repeated over and over its soft, liquid trill. Finally, after about 20 minutes, the bird descended to the ground with the cicada. Standing on the mound of freshly dug earth in front of the hole that the pillager had made, it gave the feeding call, at first in normal tones but soon rising to almost frantic intensity as the nunbird continued to call the missing nestlings. Evidently attracted by these insistent notes, a second nunbird arrived with food. This bird soon left and presently the other followed, still bearing the cicada.

Thus food was brought for the nestlings at least 24 hours after they had vanished. I have known a number of birds of other species to continue to bring food for hours or days after their nestlings were taken by predators. A pair of Golden-naped Woodpeckers did so for six days.

As in the case of the nest which had met a similar fate in the preceding year, I searched vainly for a replacement nest. This seems generally to be the way with birds of the tropical forest. If they replace a lost nest the new one is so far from the site of the old one that there is no indication that it belongs to the same pair.

Young after Leaving the Nest

On 5 May, when the young in the nest by the bluff were well feathered, a great horde of army ants (*Eciton*) swarmed over and around the mouth of the burrow, while Bicolored Antbirds, Ocellated Antthrushes, Black-faced Antthrushes, Barred Woodcreepers, and Broad-billed Motmots picked up insects and other small creatures that fled from the marauders. In the hour during which the ants were present the adult nunbirds did not appear, but I heard the nestlings' hunger call, *tuwee, tuwee,* issue from the depths of the burrow. I should have been more concerned for the young nunbirds' safety had I not, years before, watched a swarm of smaller army ants (*Labidus*) flow over the mouth of a White-whiskered Soft-wings' burrow without harming the feathered nestlings (Skutch, 1958b:229). In neither case did I notice ants entering the burrow; indeed, I have never known army ants to harm birds of any kind or age, although this is far from true of other kinds of ants, especially fire ants (*Solenopsis*).

As I had expected, the young nunbirds suffered no harm and two days later, on 7 May, the last of the brood left the nest, at the age of about 30 days. At noon on this date, guided by approach calls, I discovered adults taking food to a fledgling perching high in a forest tree about 100 feet from the burrow. On the following day, also, the attendants were carrying food into the high treetops, where they gave approach and feeding calls. After a few days the family disappeared from the vicinity, without our having seen much of the young. Sticks set upright in the burrow's mouth after the last fledgling flew indicated that it had not been reentered by 3 June, when we left La Selva. Similar observations were made at the third nest of 1968, which the young vacated about the end of April. Nunbirds do not, like certain jacamars, barbets, toucans, and woodpeckers, return to sleep in their nests; and if there is a second brood in the same burrow (which I doubt), it is started more than a month after the departure of the first.

In 1967, when we found no successful nest, we saw much more of juvenile nunbirds than we did in 1968, when two successful nests were studied. In the former year, fledglings were first seen on 28 April, about the date the young in the most advanced nest found in 1968 would have flown, had they survived. There were three of these fledglings who sat on high branches of shade trees in a cacao plantation, calling *tuwee, tuwee* in thin voices. Near them were at least four adults. The young birds, although nearly full grown, were readily distinguished from the adults by their darker body plumage, their orange-buffy (rather than white) foreheads and chins, and their whitish or light horn-colored (rather than orange-red) bills.

During the next month, or until the end of May, nunbirds continued to feed juveniles, often about the edges of the clearing in which the house stood or in neighboring cacao and banana plantations, where it was easier to watch them than in the forest. The food was usually transferred in a spectacular manner. After catching an insect, or sometimes a small lizard or frog, the adult would beat it against a branch, then hold it in its bill until a juvenile flew up and snatched it away, without alighting. Once a young bird flew a hundred feet to grab food from an adult's bill in this fashion. In coming to the attendant for food, rather than waiting for it to be brought to them, the young were, in a way, continuing the habit they had formed as tender nestlings when they came to the mouth of the burrow for their meals. They were also gaining skill for snatching insects from foliage or bark, without alighting, as they would do when older. More rarely a young nunbird would alight beside the adult to take its food, or the adult would fly up to the juvenile and deliver it. If the young were slow in coming for something that an adult had caught for them, the latter would repeat the rolling approach call until the food was taken. A young bird who received a walking-stick insect some six or eight inches long seemed embarrassed by this big meal. After a while the bird started to swallow it little by little, but a quarter of an hour elapsed before the last of the insect disappeared into its mouth.

One would suppose that young birds with such excellent control of flight as this method of taking food demanded would be quite capable of feeding themselves, but their attendants were very indulgent. Moreover, the juveniles' power of flight seemed to develop more rapidly than their knowledge of what was good to eat. Once I watched one of them, after taking food on the wing from an adult, pluck a piece of curled bark which it soon dropped from its bill. Sometimes they tried to take food from unrelated birds. As already recorded, nunbirds often associate loosely with Purple-throated Fruitcrows, black birds of about the same size. One morning in the cacao plantation, I saw a male fruitcrow catch a large green insect, which required a good deal of beating against a branch and mandibulation to make it swallowable. While the fruitcrow was engaged in this operation a young nunbird flew up beside him, as though to take the insect. The fruitcrow promptly flew to another branch, to which the nunbird pursued him, making him retreat once more with his insect, which at his next stop he gulped down.

At the end of May, when some juvenile nunbirds had been on the wing for a month or more, I looked for them to participate in the choruses that then were frequent, but I never detected a pale-billed bird in these gatherings. Not until early June did I see one catch an insect for itself.

WHITE-FRONTED NUNBIRD

Summary

White-fronted Nunbirds inhabit wet primary forests at low altitudes. In small parties of rarely more than half a dozen individuals, they straggle through the woodland canopy, whence they often make foraging excursions into neighboring clearings with scattered trees. These nunbirds were seen to take only animal food, chiefly insects, with an admixture of spiders, millipeds, and an occasional small frog or lizard.

Nunbirds are voluble, uttering a variety of expressive vocalizations, of which six are described. Loudest and most surprising is the "chorus," a bout of simultaneous shouting by from three to ten individuals who perch close together at middle heights in the forest and may perform almost continuously for 15 or 20 minutes.

At about the time that eggs are laid one member of the flock, evidently the breeding female, is fed by several others.

Nunbirds nest in a burrow 40 to 55 inches long, in sloping or nearly level ground in heavy forest. At the inner end is a chamber lined with dead leaves and around the entrance the birds arrange a low collar of twigs and leaves.

One nest held three nestlings, three others an unknown number of young, in late March, April, and May. Hatched blind and quite naked, the nestlings have prominent callous pads on their heel joints.

In each of four nests, the nestlings were attended by three or four nunbirds, all outwardly similar. In one case, four grown birds fed three young, apparently the progeny of a single female. The extra attendants were evidently nonbreeding helpers.

Unless it intended to brood, the attendant never entered the burrow with food. Even while still blind and naked the nestlings ran up the tunnel to receive their meal at its mouth, which in the longer burrows entailed a round trip of about eight or nine feet. Three nestlings becoming feathered were fed 30 times in a day. The attendants never came with more than a single insect or other small creature, held prominently in the bill.

The attendants were never seen to remove waste from the burrow, where maggots soon battened in the bed of dead leaves.

In two burrows the nestlings were destroyed by animals that dug them out. For at least 24 hours after a brood was lost the attendants continued to alight with food in front of the gaping hole and call the nestlings.

After leaving the burrow at the age of about 30 days, the young rose high into the trees. Soon they were taking food in a spectacular fashion, flying up from a distance to snatch it from an attendant's bill while on the wing. This gave practice in the nunbird's habitual manner of foraging.

After the departure of the young, neither they nor the adults returned to sleep in the burrow. Apparently, only one brood is raised.

Family RAMPHASTIDAE

CHESTNUT-MANDIBLED TOUCAN

Ramphastos swainsonii

The largest representative of its family in Middle America, the Chestnut-mandibled Toucan is about 20 to 23 inches long. Both sexes are largely black, tinged with maroon on the hindneck and upper back and with blue or green on the rest of the upper parts. The upper tail coverts are white and the under tail coverts bright red. The feathered portion of the face, the throat, and the foreneck are clear yellow, the convex posterior margin of which is bordered by a narrow line of white followed by a broader one of bright red. The greater part of the upper mandible is yellow, contrasting with the deep chestnut of its lower basal portion and the posterior part of the lower mandible. The remainder of the lower mandible is black. The brown eye is set amid bare green skin. The legs and toes are light blue. Males average larger than females, but their measurements overlap. In the color pattern of its plumage this toucan is confusingly similar to the slightly smaller Rainbow-billed or Keel-billed Toucan, the most noticeable difference being the narrow white line between the yellow and red of the breast. Their bills and voices provide the surest means of distinguishing the two species.

The Chestnut-mandibled Toucan ranges from Honduras to Colombia and Ecuador. In Central America it is found chiefly on the wetter Caribbean side, but it was formerly abundant in the heavy forests of the Pacific slope of southern Costa Rica and adjacent parts of Panama. Here it is the only large toucan, but shooting and the appalling destruction of its habitat are sadly diminishing its numbers. On the Pacific slope of extreme southern Costa Rica I have traced it upward to nearly 5,000 feet above sea level. Primarily an inhabitant of the tall rainforests, it often explores the scattered trees of neighboring clearings for edible fruits and nests of smaller birds which it pillages. In their undulating flight and habit of straggling along one behind another, Chestnut-mandibled Toucans hardly differ from Rainbow-billed Toucans.

When I arrived in the Valley of El General in the mid 1930's, the Quioros, as the Chestnut-mandibled Toucans are here called, were still abundant and fairly confiding in the presence of man, for they had not been hunted much. One day while I collected plants along a rivulet flowing down a heavily forested slope far above the valley, a party of these toucans flew about in the treetops high above me. Presently some of them descended into the lower branches to see what I was and did. While I arranged some ferns between the papers of my plant press, two of the big birds came quite close, perching no more than five yards above me, and looked on with unmistakable curiosity. One cocked his head and great bill to one side and peered down intently with a large brown eye set in an area of pale green skin in a yellow face; but the other soon became more interested in the fruit of some forest tree that it carried and began to pull it apart with the hooked tip of its bill, while holding it against a branch with a foot. These birds of the still unbroken forest were not at all shy and even when I stood up to look at them they were in no hurry to fly away.

On a sunny morning in May I watched a pair of Chestnut-mandibled Toucans bathe in a treetop pool that was doubtless well known to them. Their bathtub was a hollow in the crotch of a tall tree. First one, then the other, of the huge-billed birds flew to the narrow fork, crouched down to wet its yellow breast in the rain water that filled it, flapped its black wings as freely as the confined space would permit, then hopped up to a neighboring branch to preen and arrange its wet plumage, and dry itself in the bright sunshine that fell upon it through a gap in the high canopy of the forest.

I have never seen toucans of any kind bathe at ground level in a stream or pool, but have watched several species wash themselves in high arboreal basins. Long ago I saw a flock of Fiery-billed Araçaris bathe, one after another, in a rain-filled hollow in the upper side of a thick branch above a forest trail. More recently I watched a pair of Yellow-eared Toucanets bathe in a cavity in the side of a massive trunk at the forest's edge. First the male toucanet took his bath while his mate waited patiently on a neighboring bough. Facing outward he backed into the hollow, spilling water over the lip of the cavity whenever he immersed himself. After each wetting he rested on the rim while he vigorously preened and scratched himself, revealing the orange tufts on his sides and the yellowish under surface of his wings. When at last he left the niche, his deferential partner took her turn bathing, preening, and scratching just as he did. She continued for six or seven minutes to wet and arrange her plumage; then the pair rose higher into the trees to scratch and preen themselves a while longer.

Mated Chestnut-mandibled Toucans preen each other. Before poachers exterminated these spectacular birds in the woodland on our farm, I watched from the dooryard a pair perching side by side on a lofty branch at the forest's edge. They were alternately preening each other's plumage with the tips of their great bills. Most birds, when they perform this friendly service, rest intimately side by side, but the length of the toucans' bills made it necessary for them to remain well separated, so that at first I wondered what they were doing. Finally, one moved still farther from his companion—his way of saying "enough."

One morning before sunrise while we were looking into a nest of the Rainbow-billed Toucans at La Selva, several pairs of Mealy Parrots alighted in the top of a huge leafless tree growing nearby and gave unrestrained play to their strong, raucous voices. Soon a pair of Chestnut-mandibled Toucans, doubtless the ones whose nest cavity was close by, flew into the treetop with the noisy parrots. One held in the tip of his beak a fruit which he passed to his companion. Then he brought up from his throat or stomach, one by one, four or five other items, passing each in turn to his partner. For his reward he held up his head while the recipient of these gifts gently preened the feathers of his throat and neck with the tip of her enormous bill. In the same forest my son watched a toucan who seemed to press oil from its companion's preen gland and spread it over the plumage of the same bird.

In the foregoing account I have assumed that the donor of the food was the male and the recipient his mate, as is usual among birds, although occasionally, in several species, I have seen food passed in the reverse direction. Chapman (1929:48-50), who twice saw one Chestnut-mandibled Toucan feed another, suspected on one occasion that the generous bird was the female.

Food

Like other toucans, the Chestnut-mandibled subsists principally on fruits of forest trees. On Barro Colorado Island Chapman watched these toucans eat fruits of the wild nutmeg tree (*Virola panamensis*) and the small berries of the mangabé (*Didymopanax morototoni*). These toucans are among the many birds, ranging in size from brilliant little honeycreepers and wintering Tennessee Warblers to big, slender chachalacas, which in the dry season eagerly devour the green fruiting spikes of the guarumo (*Cecropia* spp.), a food that is, to our taste, far from savory. Where banana plantations adjoin the forest Chestnut-mandibled Toucans enter them to search for ripe fruit that the cutters have neglected.

These toucans also eat such insects as they can catch, not even disdaining very small ones. When, after an afternoon shower, the termitaries release their winged sexual generation, the air is full of weakly fluttering termites and all the feathered world, from domestic chickens to woodpeckers, feasts upon them. The toucans try hard to secure their share of the bounty. Perching on the topmost limbs of tall trees they reach far out into the air and snap their great bills with a loud clacking sound. Sometimes, at least, they capture a winged termite in this fashion, as I have seen through binoculars. From time to time the big birds fly heavily from one branch to another, apparently to reach a position where the fluttering creatures fill the air more thickly. Rarely they try to seize a termite on an aerial dart; but the flycatcher style of capturing them, adopted by nearly all birds when termites swarm, is hardly appropriate for the massive toucan.

A less pleasant aspect of the toucan's life is its addiction to the eggs and nestlings of smaller birds. The only nest of the Collared Araçari that I ever saw was situated in a thick horizontal branch of a great tree in the forest on Barro Colorado Island. The opening faced the ground, 100 feet below, and was so narrow that the slender araçaris had to squeeze through. That the hollow in the limb was deep was evident from the fact that five grown birds and at least three nestlings occupied it every night. One morning while I watched the araçaris attend their young, two Chestnut-mandibled Toucans flew into the nest tree. One of them soon discovered the nest hole, probably by hearing the voices of the nestlings. To reach the downward-facing doorway, a feat that taxed the skill of the smaller and more agile araçaris, presented an even greater challenge to the heavy toucan, but the prospect of a meal was too enticing to be neglected. Hovering clumsily below the orifice, the toucan stuck in the end of its great beak, then fell away. On its second attempt it pushed its bill somewhat farther in. Much as I doubted the would-be robber's ability to reach a nestling araçari, I wished to take no chance of losing this most interesting nest, so I emerged from my blind and tried to drive the intruder away, with ultimate success. Probably the safety of nests of the smaller toucans of lowland forests depends largely on having a doorway too strait for the bigger toucans to enter; just as the latter's success in breeding depends chiefly on having a hole in solid wood with an orifice too small for middle-sized and large arboreal mammals to enter.

Although I have never known one kind of toucan to despoil the nest of another, I have surprised Chestnut-mandibled Toucans in the act of plundering nests of several other kinds of birds. One morning, attracted by the distressful cries of the parent flycatchers and other birds nesting nearby, I arrived in time to see one of these toucans gulping down a feathered nestling of a Vermilion-crowned Flycatcher, the roof of whose domed nest the pillager had torn off. I also saw a toucan take a naked nestling of the Tropical Kingbird, but was able to drive off the marauder before it swallowed the rest of the brood. The circumstantial evidence that a Chestnut-mandibled Toucan had eaten a newly hatched Boat-billed Flycatcher was convincingly strong. All of these acts of predation occurred in my dooryard close by the forest. Farther afield I watched a toucan peer repeatedly into the unfinished nest of a Boat-billed Flycatcher, as though to see whether there were any eggs to be eaten. It snapped menacingly at the protesting flycatchers when they darted too close. I also saw a toucan pull a small, cup-shaped nest from a riverside tree and drop it into the water but I could not discover what, if anything, the nest had contained. Even small raptors are not immune from this big toucan's depredations. Laughlin (1952: 138) watched one of these birds swallow an egg of the Double-toothed Kite after driving the female from her nest.

The significance of the toucan's overgrown, colorful bill has been much discussed, without reaching any generally accepted conclusion. It it clear, however, that on the bird's marauding expeditions its great bill serves it in good stead. Such an enormous beak, even if dull-colored, would command the respect of small birds; with vivid hues to draw attention and heighten its effect, and menacing snaps if the defenders of a nest come near, it must appear a fearful weapon indeed, fit to hold aloof the parents whose nest the toucan is plundering. I have never seen even the boldest flycatcher, enraged by the violation of its nest, dare to come within range of a perching toucan's beak; even small hawks are, as we have seen, intimidated by it. But in flight the toucan appears unable to turn back its head to defend itself and its victims take advantage of this limitation to release their pent-up rage, pursuing the villain hotly and sometimes pouncing on its back, as I have seen Tropical Kingbirds and Boat-billed Flycatchers do. Yet they are careful to be at a safe distance the moment the toucan alights and can bring its bill into effective action. The Boat-bills, whose high, open nests are particularly vulnerable to toucans, have a mortal antipathy to these big plunderers and, like prudent soldiers who go forth to confront a powerful enemy before he approaches their city, fly afar to meet and harass the toucan who comes within sight of their nest.

Although they combine strength with lightness in a remarkable fashion, the hollow mandibles of toucans are not immune from accidents. Early in 1942 I saw, at Los Cusingos, a Chestnut-mandibled Toucan with the terminal inch or two missing from its upper mandible, a disfigurement visible from afar. During the next 2½ years, I met this easily recognized bird from time to time. It appeared to lead a solitary life for I always saw it alone, yet despite its impediment it seemed to keep itself well nourished. On 21 March 1968 I saw at La Selva a Chestnut-mandibled Toucan who lacked the terminal half of its lower mandible. The exposed end of its tongue was visible, through binoculars, from a distance. Yet the bird appeared to be in sound health and it was accompanied by another who was evidently its mate. Seven weeks later I again saw this toucan with a broken lower mandible, proof that it could forage well enough to stay alive—unless its mate kept it adequately supplied with food.

Voice

The far-carrying call or song of the Chestnut-mandibled Toucan is one of the characteristic sounds of unspoiled lowland forests in southern Central America. It is generally delivered from the top of a tall tree. To the devout the toucan seems to proclaim *Dios te dé* (May God give it to you), but in a more secular mood I once paraphrased it as *yó yip ă yip*. As he utters the first syllable the toucan throws his head up and back, to bring it forward and downward in a series of jerks as each succeeding syllable is enunciated. This straight up-and-down movement of the head of the singing Chestnut-mandibled Toucan is itself sufficient to distinguish it from the Rainbow-billed Toucan, who delivers his quite different song with a more complicated head movement, throwing it from side to side as well as up and down.

The Chestnut-mandibled Toucan's voice seems high-pitched and weak for so large a bird. These toucans perform more or less at all seasons but most freely in the dry early months of the year when preparing to nest. They sing much at dawn and are especially vocal as the day ends. After singing their vespers at great length, while the sun's last rays fade from the highest treetops, they fly off through the darkening forest, where most other diurnal birds have long since gone to rest, to roosting places that I could never find.

A very different sound is a hard rattle, which I long supposed to be produced mechanically, employing the resonant qualities of the great hollow upper mandible, perhaps by striking it rapidly against a branch. Finally, one morning I clearly saw a toucan open its mouth rather widely and emit a loud rattle without knocking the mandibles against each other or anything else, thereby showing unmistakably that it was a vocal sound. It evidently expresses anger, for on another occasion I heard it from a toucan defending its nest from a small mammal.

Nesting

In mid-March of 1937 I found a pair of Chestnut-mandibled Toucans preparing to incubate, or perhaps already doing so, in a noble, spreading leguminous tree (*Stryphnodendron excelsum*) growing at the edge of the forest on the mountainside above the valley of the Río Buena Vista in El General, at an altitude of 3,200 feet above sea level. Their nest cavity, about 90 feet above the ground in one of the main branches of the tree, was entered by an aperture through which the birds could barely squeeze. A single toucan slept in the hole by night and by day it was occupied part of the time.

By 24 March the toucans were certainly incubating in this inaccessible nest. If I stood below and clapped my hands loudly at almost any hour of the day, a great yellow-and-brown bill protruded through the doorway, to be followed by a pale green face with brown eyes that peered inquiringly down at me. In the gray dawn of the following day while I stood in the neighboring clearing watching a woodpeckers' nest, a sudden harsh rattle drew my attention to the toucans' tree. Looking around, I saw that the toucan had left its hole and was darting back and forth at a long-tailed animal the size of a cat, a Kinkajou (*Potos flavus*), as well as I could make out in the dim light. The animal was clambering down the trunk near the nest but after the bird had swooped close by it two or three times it turned and climbed upward. The attacks continuing, it reversed its course once more and quickly descended until lost to view behind the great trunk. After it had vanished the toucan perched near its nest and sang *Dios te dé* until the day grew brighter, when it flew off into the forest.

Unfortunately, I did not witness the beginning of this encounter; but considering the hour at which it occurred it is almost certain that the mammal's approach had surprised the toucan sleeping in its nest.

The significant fact is that when I first saw the toucan it was not within, defending its doorway with its formidable beak, as certain theories of the function of this hypertrophied organ would have it, but outside, threatening the intruder yet remaining safely beyond its reach. The bird had evidently saved its eggs for the present, for it and its mate continued to incubate through that day. But a few days later the nest was abandoned, probably because the Kinkajou had, after all, devoured the coveted eggs, perhaps also taking possession of the hole as its daytime dormitory (Skutch, 1960b).

The only other nest of the Chestnut-mandibled Toucan that I have seen or read about was approximately 30 feet up in a clean, branchless trunk of a large, living tree of *Pentaclethra macroloba* on a forested ridge at La Selva. The cavity had evidently been created by the decay of a knot rather than carved by a bird. This nest was only 50 feet from the simultaneously occupied Rainbow-billed Toucans' nest that we studied in some detail (Skutch, 1971a). Since Van Tyne (1929) had long before published an excellent account of the latter species, I should have preferred to study its larger neighbor; but the higher nest of the Chestnut-mandibled Toucan was on the underside of a leaning trunk and could not safely be reached with any equipment available to us.

Although the Chestnut-mandibled Toucans had been interested in their hole at least as early as the second week of April 1967, they appeared to be incubating, or possibly brooding recently hatched nestlings, when we left La Selva in early June. In the following year they seemed to be incubating in the same hole on 20 March, and by 7 April they were taking food to it. But apparently some mishap befell the nestlings, for in mid-May these toucans were incubating again in the same hole, or preparing to do so. On 2, 9, and 17 May a single toucan flew out at dawn after having passed the night in the hole. Whatever the detailed history of this nest cavity might have been, the toucans' persisting attachment to it was evident.

It was also clear that there was no serious antagonism between these toucans and the closely related Rainbow-bills, who in both seasons raised a nestling only 50 feet away. Once, indeed, a parent Rainbow-bill, delaying noisily to take food to its nestling, hastily swallowed its fruit and departed as one of its larger neighbors flew toward it, but I saw no repetition of such behavior. On another occasion, when both parent Rainbow-bills were complaining loudly in the treetops while we inspected their nest, a Chestnut-mandibled Toucan joined them and added its more dignified *Dios te dé* to their croaks.

Summary

Primarily an inhabitant of heavy forest, from sea level up to nearly 5,000 feet, the Chestnut-mandibled Toucan often wanders through neighboring clearings with scattered trees in search of food. Where it has not been persecuted it sometimes manifests unmistakable curiosity about human activities.

Like other toucans it bathes in rain-filled hollows in branches and crotches, high in trees.

Mated toucans alternately preen each other's plumage, and they feed each other.

These toucans subsist largely upon fruits, supplemented by insects and small vertebrates. When all the avian world is feasting on the flying sexual generation of termites, the toucans try clumsily to catch them, too. They devour the eggs and nestlings of other birds, including even small hawks. The large, brightly colored bill serves the toucan by intimidating the parent birds whose progeny it takes.

The hollow mandibles of toucans are occasionally broken, yet the birds manage to survive for long periods despite this handicap.

The toucan's song, a phrase of four syllables sometimes rendered *Dios te dé,* is delivered with a pronounced up-and-down movement of the head. It is heard chiefly at dawn and, even more, as the day ends. A castanet-like rattle is produced vocally, as in the Rainbow-billed Toucan.

In Costa Rica the nesting season extends from March to at least June. Two nests were in cavities, evidently resulting from decay, about 90 and 30 feet up in living trees. The contents of this toucan's nest seem never to have been described. When a small mammal approached its nest an incubating toucan emerged to threaten the animal from outside, showing that the great bill is not, as has been suggested, employed to defend the doorway while the parent remains within.

In two consecutive years Chestnut-mandibled Toucans and Rainbow-billed Toucans nested only 50 feet apart, without apparent antagonism.

Family FORMICARIIDAE

FASCIATED ANTSHRIKE

Cymbilaimus lineatus

The Fasciated Antshrike is a stout, strong-billed antbird about 6½ inches long. The male has a black crown and is nearly everywhere else barred with black and white. On his upper plumage the white bars are narrow and the black interspaces broad, but below the white and black lines are of about equal width. The female has a buffy forehead and cinnamon-rufous crown, both more or less barred with black. The rest of her upper parts are blackish, broadly barred with buff. The sides of her head and neck and all her under parts are pale buff or tawny buff narrowly barred with black. In both sexes the abruptly hooked upper mandible is black, the lower mandible pale bluish gray. The eyes are bright red, the legs and toes leaden blue.

This antshrike ranges from Honduras to Peru, the Guianas, and northern Brazil. In Costa Rica, where it is confined to the Caribbean lowlands and foothills, it is not known to occur above 2,000 feet (Carriker, 1910:601) but in Venezuela it extends upward to 4,600 feet (Phelps and Phelps, Jr., 1963:79). Fasciated Antshrikes live obscurely amid the heaviest foliage at the forest's edge, in older second growth and the overgrown margins of streams, where they forage at no great height. Carriker found them in patches of wild cane along the Río Sixaola. They subsist largely, if not wholly, upon insects, spiders, and other invertebrates, which they catch amid the foliage. Their presence would rarely be noticed were it not for their voice. One call consists of soft notes which resemble those of the Black-throated Trogon but are delivered in longer series. This is evidently the song that Chapman (1929:277) called a "soothing little chant" and Eisenmann (1952:33) described as "an easily imitated series of whistles, all on one note, uttered at a rate of about two a second, and continuously repeated for 8 to 10 seconds, *cü, cü, cü, cü, cü* or *cwee, cwee, cwee,* etc." This utterance, which carries far, is very ventriloquial, failing to reveal the exact location of the antshrike lurking amid exuberant verdure. A quite different vocalization is a rattle, which may be continued for minutes together, and when loudest resembles the rattle of a large kingfisher. The rattle may be punctuated by low, soft notes or followed by a little whine. The alarm note, according to Eisenmann, is a nasal *wañurk* (or *wanyurk*).

Nesting

The only nest of the Fasciated Antshrike of which I have information was found at La Selva by my boy Edwin on 19 April 1967. It was at the edge of a cacao plantation about 25 feet up in a tree above a stream flowing between high banks covered with dense bushy growth and scattered trees. The ample open cup was suspended by its rim between the slender arms of a horizontal fork. The fabric, so open that much light passed through, was composed largely of dark fibers. A creeping polypody had been woven into it and the small green leaves of this epiphytic fern covered much of the outer surface of the cup, making it difficult to detect amid the foliage clustered around it. We could not by any means, short of cutting down the tree, see the eggs that the antshrikes were incubating, but probably there were two, as is usual in the Formicariidae.

As in all other antbirds, as far as we know, the female incubated by night and the two sexes took turns by day. We watched them throughout the day of 20 April, through the morning of 27 April, and the afternoon of 30 April. At 05:40 on 20 April we found the female in the nest. She continued to sit quietly until 06:59, when she left as her mate came through the cacao plantation, announcing his approach with a rapid loud rattle, punctuated at intervals by soft, contrasting notes. He brought a billful of fibers which he added to the nest. Then he settled on the eggs, continuing for two minutes more to rattle with his lower mandible rapidly vibrating. After sitting in silence for 1¾ hours, he began to call, repeating over and over a phrase consisting of three to five soft, rather plaintive notes. Three minutes later, still calling, he left the nest to forage amid the neighboring foliage. After an absence of 14 minutes, passed near the nest, he returned, rattling, and incubated for another hour, during which from time to time he called softly. Finally at 10:05 he went off to a distance and for the next 70 minutes the nest remained unattended. Then the female came, calling, and incubated for 45 minutes, or until noon, when she was replaced by the male. For the next 4¼ hours he alone took charge of the nest but he did not sit continuously. On the contrary, he left from time to time to search for insects near the nest, then returned for another session on the eggs. This continued until 16:26 when the female returned, just as a heavy downpour began, and remained on the nest until evening (see Table 2).

Thus, on this day the female took only one session, lasting 45 minutes, between her first morning departure at 06:59 and her return for the night at 16:26. The male took seven sessions, ranging from seven to 107 minutes and totalling 372 minutes. The eggs were left exposed for eight intervals, ranging from four to 70 minutes and totalling 150 minutes. Or, if we consider the diurnal period extending from 05:45 to 17:45, the female incubated for a total of 198 minutes and the other totals remain the same.

On the rainy morning of 27 April the female continued on the nest until 06:34, when her mate announced his approach with rattling notes. He came into view eating a large insect and promptly settled

TABLE 2
INCUBATION BY A PAIR OF FASCIATED ANTSHRIKES

20 April 1967, 06:59-16:26 (Intervals in minutes)				27 & 30 April 1967, 06:34-15:58 (Intervals in minutes)		
Male	Female	Neither		Male	Female	Neither
			a.m.			
		4		197		2
107		14			127	
61		70				
	45					
			p.m.			
106		16		43		10
7		14		24		14
37		14		17		14
12		7		10		2
42		11			64	
				40		
372	45	150		331	191	42

Average of 13 sessions by male, 54.1 minutes; range 7-197 minutes.
Average of 3 sessions by female, 78.7 minutes; range 45-127 minutes.
Average of 13 intervals of neglect, 14.8 minutes; range 2-70 minutes.

on the eggs, where he stayed until he heard his mate coming at 09:51. Two minutes later she entered the nest and was still sitting when observation was suspended at noon. On the afternoon of 30 April the male, who was in charge of the nest most of the time, again sat less constantly than in the morning, leaving the nest from time to time to hunt insects nearby. The female took one session of 64 minutes and 40 minutes later, or at 15:58, returned, to remain until nightfall. Combining the records made in the morning and afternoon watches, which together covered all hours of the day, the male antshrike took six sessions that ranged from 10 to 197 minutes and totalled 331 minutes, the female two sessions of 127 and 64 minutes, and the eggs were unattended for five intervals ranging from two to 14 minutes and totalling 42 minutes. Or, if we consider the whole diurnal period from 05:45 to 17:45, the female sat for 347 minutes and the other totals remain unchanged.

The female antshrike always sat facing outward from the fork that supported the nest. The male incubated with his head either inward or outward, shifting occasionally from one orientation to the other in the course of a session, and more rarely he sat sideward. Both were quite confident in their high nest and did not object to our watching them while sitting unconcealed at the edge of the cacao plantation. One morning I tried to view the eggs in a mirror attached to the end of a slender young tree, but I had too little control over such a long, heavy, and limber pole and did not succeed. The female, who was incubating at the time, sat until the mirror shook the leaves close beside her. Then she hopped through the foliage of the supporting branch, repeating low, soft, mournful monosyllables. As soon as the mirror was lowered she returned to her eggs while I stood almost below.

When she came to take charge of the nest at 15:58 on 30 April the female brought a small insect. Perching beside the nest she lowered her head into it; the insect disappeared. I thought that the nestlings had hatched but careful watching did not reveal their presence until three days later, when feeding became regular. What I witnessed on 30 April was an instance of anticipatory food-bringing, such as occurs occasionally in many kinds of birds (Skutch, 1953:10-17). Indeed, both sexes of the antshrikes had a tendency to carry something as they approached to take their turns at incubation. Sometimes one would be eating an insect as it came into view and twice I saw the male bring fibers to the nest. Such continuance of building is common enough in birds that construct big, bulky nests but rare among those whose nests are as compact and neat as the Fasciated Antshrikes'.

On 3 May both parents were feeding nestlings and eating their droppings. But before we could study the details of parental care the nest was empty, having evidently been pillaged by some predator. We searched in vain for a replacement.

Summary

The Fasciated Antshrike inhabits dense vegetation at the forest's edge, along streams, and in long-abandoned clearings, all in humid lowlands. It appears to subsist largely upon insects and other invertebrates that it catches amid heavy foliage.

The call, or song, is a long series of soft, mellow notes, which resemble those of the Black-throated Trogon and are very ventriloquial, failing to reveal the bird's location. It also utters a rattle, similar to that of a large kingfisher.

The single nest of which I have information was situated about 25 feet up in a tree beside a stream. Suspended by its rim between the arms of a horizontal fork, the ample open cup was composed largely of dark fibers. The eggs could not be seen.

The female incubated by night and by day the sexes alternated on the eggs, with the male sitting much more than his mate. Diurnal sessions by both sexes ranged from seven to 197 minutes in length. Intervals when neither parent covered the eggs ranged from two to 70 minutes.

Before the eggs hatched the female brought food to the nest, anticipating the birth of the nestlings.

Both parents fed and brooded the nestlings, which disappeared when only a few days old.

STREAKED-CROWNED ANTVIREO

Dysithamnus striaticeps

A chubby little bird with a conspicuously short tail, the Streaked-crowned Antvireo is only four inches long. In the male the top of the head is slate-gray, broadly streaked with black. The upper surface of the body is olive or grayish olive, the tail russet-olive or olive-brown. The remiges and their greater coverts are olive, the latter tipped with brownish white, forming a narrow wing band. The middle and lesser coverts are black, conspicuously spotted with white. The cheeks are gray, flecked with dusky. The chin, throat, and chest are white, boldly streaked with gray. The abdomen is white, passing into buff on the under tail coverts, and the sides and flanks are buffy olive. The female differs chiefly in having the pileum pale cinnamon or russet with dusky streaks, the spots on the wing coverts buffy instead of white, and the cheeks and under parts buffy with more or less distinct darker streaks on the cheeks and chest. In both sexes the bill and eyes are dark and the feet are gray.

The Streaked-crowned Antvireo is found only on the Caribbean slope of Honduras, Nicaragua, Costa Rica, and adjacent parts of Panama. Carriker (1910:607) wrote that in Costa Rica it "is confined entirely to the Caribbean foothills, is rarely seen below 800 feet and is most abundant at about 1,000 to 1,500 feet, especially in north-eastern Costa Rica." Slud (1960:100), however, stated that it was common at La Selva, where most of the land lies between 300 and 500 feet. In this locality I found this antvireo chiefly in the mixed flocks of small birds that roamed through the lower levels of the heavy forest, restlessly foraging between about six and 30 feet above the ground. The noisiest and most conspicuous member of these flocks was nearly always the Tawny-crowned Greenlet, associated with which were such other species as the Wedge-billed Woodcreeper, Eye-ringed Flatbill, and White-flanked Antwren. Among the least prominent members of these mixed parties were the Streaked-crowned Antvireos, so quiet and retiring that it was difficult to detect more than one amid the dense, dimly lighted undergrowth. There was never a band of "from five to ten," such as Carriker mentioned; indeed, ant-birds of any kind are rarely so gregarious, unless they are following army ants, as these antvireos seldom, if ever, do. Streaked-crowned Antvireos subsist largely, perhaps wholly, on insects and spiders gleaned from the foliage. Their call, or song, is a little wooden roll terminating in an emphasized note. They complain with a low *pur-r-r,* interspersed with slight, mournful monosyllables.

NESTING

A nest found at La Selva on 5 May 1967 was 14 feet above the ground in a horizontal fork of a slender lateral branch of a small tree growing in the forest. The nearly hemispherical cup, measuring three inches in diameter by 2½ inches in height, was attached by its rim to the thin arms of the fork. The open fabric, composed of dark fibers, was lightly covered on the outside with green moss, long streamers of which hung below the nest. The two whitish eggs were blotched and flecked all over, very heavily on the thicker end, with chocolate. There was a tendency for the heavier marks to be elongated in the direction of the long axis of the egg. These eggs measured 22.4 by 16.5 and 22.7 by 16.6 mm.

Carriker (1910:607) recorded a nest of this antvireo, of similar construction, that was "suspended from a horizontal fork of a small tree near the edge of the forest, about five feet from the ground." On 12 May 1905 it contained two well-incubated eggs that resembled those already described and measured 21 by 15 and 20.5 by 15.5 mm.

For such small birds, these antvireos incubated for surprisingly long intervals. On 6 May, the day after we discovered the nest, I arrived at 08:45 and found the male covering the eggs. He was still there when the female arrived with a mixed flock of small birds that was passing by. She seemed about to go to the nest, but seeing me watching some distance away, she flew off. The male was still on the nest when I left at 11:45, having sat continuously for at least three hours.

After an incubation period of no less than 14 days, two nestlings hatched on 19 May. When these nestlings were three days old I watched them from 08:00 to 10:30, mostly beneath a hard rain. Two minutes after I arrived the female delivered a large green insect, swallowed a dropping, and covered the nestlings. After sitting for 40 minutes she left as her mate came to feed and brood. He sat for 80 minutes, leaving when the female arrived, again as the mixed flock making its rounds of the undergrowth passed near the nest. After giving a large insect to a nestling, she brooded.

On 25 May, when the two nestlings were six days old, we watched from 06:40 to 11:40. The male brought food four times and the female three times, making seven meals in five hours. On each visit a single item, usually a fairly large insect, was brought, held conspicuously in the parent's bill. Each parent brooded twice, the male for a total of 124 minutes and the female for 42 minutes. Both carried away fecal sacs in their bills.

At 07:30 on 30 May a stubby-tailed fledgling was perching about one foot from the nest, in which the other nestling remained. Neither changed its position in the next hour, during which both parents brought them food. When the female noticed me watching she flitted around for many minutes, uttering, with closed bill and swelling throat, a low purring sound, punctuated by little plaintive notes. Except for this complaint the parents were always discreetly silent when in view of their nest. By mid-afternoon of the same day the nest was empty and both young antwrens had vanished after a nestling period of 11 days.

Summary

The Streaked-crowned Antvireo inhabits wet lowland forest where it wanders through the lower levels in mixed parties of small birds, hunting its insect food amid the foliage.

Its song, or call, is a little wooden rattle with an emphasized final note. It complains with a low *pur-r-r* punctuated by slight, mournful monosyllables.

The nest is a hemispherical cup hung, vireo-like, between the thin arms of a fork, about five to 14 feet up. It is composed of dark fibers, with some green moss on the outside. The two whitish, speckled eggs are incubated by both parents and hatch in 14 or more days. Both parents brood and feed the nestlings, which are feathered and leave the nest when 11 days old.

Family RHINOCRYPTIDAE

UNICOLORED TAPACULO

Scytalopus unicolor

In the race of the Unicolored Tapaculo that lives high on the eastern slopes of the Ecuadorian Andes, both sexes are black nearly everywhere, including the eyes and the short rather stout bill. The legs and toes of these six-inch birds are blackish. My acquaintance with the Unicolored Tapaculo was limited to the discovery of a nest and a meeting with a parent followed by a fledgling. Since so little is known about the habits of this genus, or indeed about the small, largely South American family to which it belongs, I give here what little I succeeded in learning about this elusive bird.

On the morning of 8 October 1939 I climbed a trail that from the town of Baños, Ecuador, led upward through a great *barranco* or canyon near the foot of Volcán Tungurahua toward its snowy summit far above. While passing through a thicket in the bottom of the gorge I happened to espy, near the ground among the bushes below the trail, a small black bird with an uptilted tail and a short black bill full of insects. Halting immediately I kept my eyes fastened on the strange bird. After fidgeting for a few moments it flew across the path and vanished amid the delicate selaginella that covered the bank to my right with a dense verdant tapestry, from beneath which issued the sizzling cries of nestlings receiving a meal.

I waited a little longer and soon this bird or its mate again took insects to the bank. Since these two parental visits showed me the approximate location of their offspring, I went to the bank, hopeful of finding a nest. But prolonged careful searching failed to disclose either nest or young. Completely baffled, I walked away a short distance, so that the tapaculos might continue to carry food to their young and reveal the exact spot where they were hidden. Presently both parents returned bringing food and after a little hesitation one of them delivered it, while I stood near enough to note precisely where the bird disappeared beneath the fronds of selaginella. Even with this guidance I was long in finding the cunningly concealed nest; without help from the parents themselves I might have searched all day without disclosing it.

At this point my aneroid barometer read 7,400 feet. The bank covered with selaginella, amid which grew a few small ferns and grasses, was eight feet high, rising vertically above a sloping base. The tapaculos' nest was situated where the upright wall joined the gentler slope, in a niche at the back of a little recess well hidden by the sprays of selaginella that draped over it. The niche was roughly globular, about three inches in diameter, and completely lined on the bottom, sides, back, and top with fine, richly branched, capillary, black material, chiefly stems of a moss or liverwort from which the tiny leaves had fallen. A few light-colored bast fibers formed an inner lining on the bottom. The round entrance to this cozy globe was about 1¾ inches in diameter. Considering how closely the nest fitted the niche, I think it probable that the latter had been dug by the birds themselves.

In this snug chamber were two nestlings with expanding plumage. That on the back promised to be brown, finely barred with black. The ventral feathers were buffy with narrow dusky tips. The expanded ends of the remiges were dusky.

My elation on finding the only nest of any member of the tapaculo family that I have ever seen was dampened by the fact that it was so far advanced, and so far from my residence, that I could not make a thorough study of it. But after completing my notes I sat unconcealed, only three yards away, to learn what I could about the parents' habits. Both fed the nestlings, bringing them billfuls of winged insects, larvae, pupae, spiders, and the like. Approaching the nest they hopped through the dense undergrowth and over the ground, sometimes vanishing beneath the lowest layer of vegetation and the litter of decaying leaves and fallen twigs. The birds held their tails strongly uptilted. The tail feathers of one of them were badly frayed from so much movement through dense undergrowth. When I first found these tapaculos they uttered low, throaty notes while they hopped about amid the bushes below the path, hesitating to go to their nest. After they had become accustomed to their watcher they sometimes delivered a rapid sequence of very low notes as they approached the niche in the bank. They were soon so confident that one of them flitted through the shrubbery only a yard above my head, with food in its bill, while I again examined its nest.

Earlier that same morning, while I toiled up a particularly steep stretch of the stony trail, I glimpsed among the bushes beside it a Unicolored Tapaculo followed by a brownish bird of nearly equal size, evidently a full-grown fledgling.

Summary

A nest of the Unicolored Tapaculo, found in October at about 7,500 feet on the eastern slope of the Ecuadorian Andes, was a globular structure of richly branched, black, capillary material that fitted snugly into a niche in an earthen bank covered with selaginella and fern. It was so well concealed that even with guidance by the parents it was difficult to find.

Both parents fed the two nestlings, hopping over the ground and disappearing beneath litter as they approached with billfuls of winged insects, larvae, and spiders.

Family TYRANNIDAE

LONG-TAILED FLYCATCHER

Colonia colonus

A small blackish bird with a disproportionately long tail, the Long-tailed Flycatcher is difficult to confuse with any other species within its extensive range from southern Honduras to Paraguay and southern Brazil. The male measures about nine inches in length, but his attenuated central rectrices take up six or more of these inches. The shorter-tailed female is about 7¾ inches in total length. In size and coloration of the body the two sexes differ little. Their black plumage is relieved by a sooty gray cap that fades to white on the forehead and above the eyes and ears. In the northern form, *leuconotus,* with which this account deals, the center of the back is broadly streaked with pale gray or dull white and the rump is grayish white. The central tail feathers present a variation of the racquet shape found in a few other birds: narrowed to scarcely more than the shafts midway of their length, they gradually expand backward into long, slender ribbons. In both sexes the bill, eyes, and feet are dark.

An inhabitant of humid lowlands, the Long-tailed Flycatcher hardly ascends to 2,000 feet on the Caribbean slope of Costa Rica. Its favorite habitat is a recent clearing in the forest where a few dying or dead trees, riddled with woodpeckers' holes, stand gaunt and forlorn above plantings of cacao or bananas, pasture grass, or the riotous second-growth that blankets an abandoned milpa or provision plot. Often, too, these flycatchers are seen in the trees that line the banks of waterways. Perching conspicuously on high, exposed stubs, male and female usually not far apart, the brisk little birds make innumerable sallies into the air to seize insects that are usually too small to be detected from the ground. Their long tails, especially that of the male, whistle in the breeze as they dart and twist and turn with admirable skill. Insects seem to constitute their whole diet; I have not seen them eat berries, as many flycatchers do.

With high, thin *pee*'s, continually repeated, the pair keep in contact while they flycatch from neighboring dead trees. Occasionally in the nesting season the male delivers longer phrases in a deeper, slightly "mournful" voice. One of these calls sounds like *we-e-e we we we* (the *e-e* representing a slight quaver). Other versions are *bee we-e we we-e* and *be-e-e we-e-e.* One must be attentive to hear these slight utterances. These more complex phrases suggest that, like many other members of their family, male Long-tailed Flycatchers have a special dawn song; but if so, I have failed to hear it in both Ecuador and Costa Rica.

Sleeping

On two occasions I tried to learn whether Long-tailed Flycatchers use holes in trees as dormitories as well as for nesting. One evening at the end of April I stood in a cacao plantation watching a great dead surá tree (*Terminalia lucida*), in one of whose many cavities these flycatchers were building a nest. The Collared Araçaris and Black-cheeked Woodpeckers retired into their sleeping holes high in the tree and a family of Banded-backed Wrens entered their nest hidden in a clump of orchids and other epiphytes far up on the barkless trunk; but the flycatchers continued to dart out from their high perches for insects. After a while the female vanished, probably having entered the cavity in which she was building. The male remained standing on the end of his stub. When a belated Black-cheeked Woodpecker arrived to sleep in a hole directly above that which the flycatchers had chosen for their nest, he darted menacingly at the bigger bird but could not keep it out. In the failing light the long-tailed bird continued to catch insects high in the air, and several times he flew up in front of the hole where the woodpecker was ensconced. A Chestnut-mandibled Toucan alighted on top of the dead tree, then continued onward in the twilight to its roost, wherever it was. The late-retiring wood-creepers sounded their parting notes in the dusk but still the little flycatcher remained silhouetted against the darkening sky. Finally, after the crepuscular Pauraque began to sing, he darted away and vanished over the treetops.

A month later when this pair of flycatchers, having lost their first nest, was preparing to renest in a smaller dead trunk also with many holes, I watched again as night descended. For a long while both flycatchers continued to make spectacular aerial sallies for insects. When a female Black-cheeked Woodpecker arrived to sleep in the trunk both flew at her without making her deviate from her course. Then the female flycatcher retired into the cavity where she was preparing to lay her eggs. The male went to peer into a slightly higher hole which communicated with the one where she rested. While the female stayed inside the male continued to perch on a high stub, at first catching insects, then outlined motionless against the fading sky. When a Collared Araçari retired into an old woodpecker hole well above that which the female flycatcher occupied, her mate circled excitedly around and around the trunk, his long tail streaming. Soon he returned to his favorite stub and stood there immobile. Swifts vanished from the sky, parrots and woodcreepers fell silent. Finally, at 16:27 when it was nearly dark and the Pauraques had awakened to sing, the male flycatcher suddenly flew across to the neighboring forest and was lost to view—one of the last of the diurnal birds to go to roost.

From these observations I conclude that the female Long-tailed Flycatcher sleeps in her nest hole before she begins to incubate, as many other birds that breed in holes or closed nests do, but the male roosts amid the foliage. To sleep in a confined space would be prejudicial to the long tail that is his chief adornment. This mode of sleeping would probably be betrayed by a persisting curvature of his central rectrices, which I have never noticed.

Nesting

In an earlier work (Skutch, 1960a) I gave a brief account of this flycatcher based on observations made at an inaccessible nest in the eastern foothills of the Andes in Ecuador. In two seasons at La Selva I saw five other nests, likewise all beyond reach. The lowest was about 25 feet up in a slender dead trunk too rotten to be climbed; the others ranged from about 70 to 125 feet above the ground. Although the inaccessibility of all these nests limited what could be learned about them, they enabled me to fill some gaps in our knowledge of these delightful birds.

On 23 and 24 April 1967 I watched a pair building high up in the big dead surá tree, in a cavity close below another in which a pair of Black-crowned Tityras were feeding nestlings. Both the male and female flycatchers built. On the first morning they took nearly equal shares. On the second morning the male at first brought material more often than his mate. After she warmed up to the work he was content to perch upon his high stub, flying out from time to time to catch an insect, while she carried billful after billful into the hole. Was he suggesting by his early trips with material that she get to work? After she had built a while alone he brought a few more pieces to the nest. All the material, consisting chiefly of fine rachises, petioles, and fragments of epiphytes, was gathered from high in the trees rather than the ground. On several occasions when the female dropped material while trying to pass it through the doorway, the male darted down, caught the falling piece in mid-air, and took it into the hole.

The fall of a rotten branch containing a Long-tailed Flycatcher's nest, in a hole carved by Black-cheeked Woodpeckers, gave me an opportunity to examine its structure. In the bottom of the cavity I found a thick mat composed almost wholly of the slender, curving, dark-colored rachises of a compound leaf. Most were between 2½ and four inches long.

Incubation is performed by the female alone. I spent the clear morning of 9 June making a record of one incubating female's movements. She was one of the most restless sitters I have ever watched. In five hours she took 30 sessions ranging from one to eight minutes in length and averaging 4.5 minutes. An equal number of recesses varied from two to 13 minutes and averaged 5.5 minutes. She was in her nest only 44.5 per cent of the five hours. No other bird that I have watched incubate has taken such consistently short sessions nor made such a low average; but a few, including a White-winged Becard and a Torrent

Flycatcher, have sat with about the same low constancy (Skutch, 1962b, Table 2). Only the Black-headed Tody-Flycatcher has incubated with a lower constancy, but some of her sessions were much longer than any that I timed for the Long-tailed Flycatcher.

The cavity in which this flycatcher incubated had two round openings, made by woodpeckers, which admitted much light. The lower aperture was smaller and through this she always entered and left but her mate sometimes looked into the nest through the upper opening, large enough to admit a Pale-billed Woodpecker. He made these periodic inspections chiefly when his mate was absent and the eggs were visible. While the female incubated I could see her long central tail feathers, which rose above her back with the tips bent forward beneath the thick upper edge of the smaller doorway. Had these two feathers been a little longer, they would have projected into the outer air like the two long plumes in the train of an incubating male Quetzal. These rectrices had acquired a pronounced upward curvature which persisted while the female flycatcher was in the outer air.

Often the flycatcher seemed to be lured from her nest by an insect flying in front of her doorway. Darting out she would catch the insect, then proceed to her preferred perch on an exposed twig of a neighboring tree from which she pursued her flycatching. Her return flight to the nest was usually the continuation of a long dart on which she caught a volitant insect. Her mate spent most of the morning perching on a stub near the nest, making frequent long sallies for insects, constantly repeating high, thin notes and occasionally voicing the longer "mournful" phrases already mentioned.

After the nestlings hatched the male helped to feed them. On 5 May I noticed a fledgling from an earlier nest still fed by its parents. Its plumage was wholly blackish and it lacked elongated tail feathers.

Summary

In humid lowlands, Long-tailed Flycatchers frequent wooded banks of waterways and clearings beside the forest, especially those with standing dead trees. They subsist largely, if not wholly, on small insects captured on aerial sallies.

A female slept in her nest before incubation began but her mate flew into the forest, evidently to roost amid the foliage. He was one of the last diurnal birds to retire.

These flycatchers nest in old woodpeckers' holes and other cavities in trees, often at a great height and rarely as low as 25 feet. Both sexes carry into the hole materials gathered from trees rather than from the ground, chiefly slender, curving rachises.

Only the female incubates, most inconstantly. One female incubated for only 44.5 per cent of a clear morning; her longest session lasted only eight minutes.

Both sexes feed the young.

PIRATIC FLYCATCHER

Legatus leucophaius

The most obstreperous member of a large and generally well-behaved family, the Piratic Flycatcher is a bird of somewhat dingy aspect slightly under six inches long. In both sexes the sooty brown crown, with a concealed patch of yellow in its center, is bordered by bands of dull white passing above each eye to the hindhead. The rest of the upper plumage is deep grayish brown. There are inconspicuous whitish wing bars and light margins on the remiges. The lower cheeks, chin, and throat are white. The pale sulphur-yellow breast and abdomen are streaked with dusky. The short, broad bill is black and the legs are dark gray.

This flycatcher that causes so much disturbance in the avian world ranges latitudinally from southern Mexico to northern Argentina, and altitudinally from sea level up to about 5,500 feet, at least in Costa Rica. North of Panama it appears to be only a summer resident, arriving from the south in late January or February, nesting, and leaving in August or September. It inhabits open country with scattered trees rather than heavy forest. In an earlier work (Skutch, 1960a: 451-464) I gave an account of its habits, telling how it captures the covered nests of a variety of other birds by harassing them and, more effectively, throwing out their eggs. The purpose of the present account is to record some more recent observations and interpretations.

Two New Hosts

In the above-mentioned account I listed nine species of birds whose closed nests are known to be captured and used for breeding by the Piratic Flycatcher. To this I can now add two more, the Rufous-fronted Thornbird and the Scarlet-rumped Cacique. In a long, pensile pouch built by the latter about 70 feet above the shore of the Río Puerto Viejo in the Sarapiquí lowlands of Costa Rica, a pair of Piratic Flycatchers were feeding nestlings in mid-April of 1968. The occupied pouch was one of two similar nests (the other vacant) that hung close beside a large, top-shaped, carton vespiary.

Near Pirapira in the state of Carabobo in north-central Venezuela, only one of the many nests of the Rufous-fronted Thornbird that I had under observation in 1966 was occupied by Piratic Flycatchers. This was of special interest as it showed how the persecuted species may continue for a long while to live in close contact with its persecutor. The nest in question hung about 50 feet up in a large *Cordia* tree in a pasture. When found on 26 March the bulky structure of interlaced sticks contained two completed chambers, each with its separate open-

FIG. 14. Nest of Rufous-fronted Thornbird containing several chambers. Nests like this are sometimes occupied by Piratic Flycatchers, Blue Tanagers, and various other birds. Near Pirapira, Estado de Carabobo, Venezuela, May 1966.

ing to the exterior, and the indefatigable thornbirds were carrying up sticks to build a third compartment, on top of the second, which in turn was above the first and lowest. At almost any hour of the day, until late in the evening, the pair of Pirates were to be found on or near the thornbirds' nest, calling insolently *pee-e-e-e* and *pee-de-de-de*, much as their kind does in Central America. As though stimulated to greater activity by the presence of these intruders, the thornbirds worked unusually hard, two of them often continuing to carry up sticks until sunset. Much of the time each kind of bird ignored the other, but occasionally a Pirate pursued a thornbird as it flew upward with a stick or downward to find another. More rarely a thornbird, tried beyond endurance, flew angrily at a Pirate, making it flee, while the flycatchers' mate, as often happens with these adept at persecution, pursued the pursuer.

Thornbirds use their nests as dormitories as well as for rearing their young, and in March three of them slept in the structure that was coveted by the flycatchers. In the evenings one of the latter, whose sex I could not tell, lingered by the nest into the twilight in an attempt to keep the thornbirds out, both by attacking them and by blocking the doorway of the chamber in which they lodged. But thornbirds have great determination, as is needed to build nests as large as theirs, and are not easily deterred from their purposes. Until their newest chamber was ready for occupancy they retired into the middle one, despite all the flycatchers' efforts to keep them out.

At 18:15 o'clock on 5 April I found the Pirates at the nest while the thornbirds carried up sticks. As it grew dark three of the latter ascended inconspicuously through the foliage of the *Cordia* tree, and coming over the top of the nest they entered the unfinished upper compartment without being molested by the flycatchers, perhaps without being noticed by them. Both flycatchers remained clinging to the nest for many minutes after the thornbirds entered. At last, in the waning light, one flew off while the other still clung to the doorway of the central chamber that they claimed. Of a sudden, in the dusk, a thornbird emerged from the topmost compartment, came down over the side of the nest and pushed into the middle compartment right in front of the flycatcher, who spread its wings over the doorway in a vain attempt to keep the thornbird out. The other two thornbirds followed, likewise forcing their way into the chamber in front of the flycatcher. Still the Pirate remained clinging to the doorway in the fading light. Soon, even through my 8 x 40 binoculars, I could distinguish only the flycatcher's light head markings, and presently these also became difficult to see: the bird faded from vision. Finally, when it was nearly dark, I thought I saw the Pirate fly away. It was then 19:15.

As was to be expected, the flycatchers, who subsisted on berries and aerial insects, nested earlier than the thornbirds, who hunt for small creatures in the ground litter, where they become more abundant after the rains return. By mid-April, when the thornbirds were sleeping in the newly finished top chamber, the female flycatcher was incubating in the compartment below them. By late April the flycatchers were feeding nestlings, which apparently made a safe departure. By early June the flycatchers were incubating in the middle compartment for the second time while the thornbirds were at last attending their first set of eggs in the top one. From at least 28 June to 6 July both pairs busily fed their respective broods in contiguous chambers. Occasionally a Pirate perfunctorily chased a thornbird, but mostly each of the four parents minded its own business. On the evening of 18 July two thornbirds retired into the top compartment, but the Piratic Flycatchers were nowhere to be found. Since thornbird fledglings would have returned to sleep in the nest, it was evident that the parent thornbirds had somehow lost their brood. Piratic Flycatchers do not use their stolen nests as dormitories and I could draw no conclusion from their failure to appear.

Destruction of Nestlings

In my earlier account I told how Piratic Flycatchers gained possession of a nest of a Gray-capped Flycatcher by throwing out the builder's eggs; and how, when safely installed and incubating in a stolen nest, they may wantonly remove an egg from a neighboring covered structure. But I knew of no instance of the molestation of nestlings of another species by the Piratic Flycatcher. Soon after my book was published, however, an instance of this came to my attention. In May of 1960 a blackish, retort-shaped nest of the Sulphury Flatbill hung ten feet up from a slender twig of an orange tree in the pasture in front of our house. As I passed by early on 7 May a pair of Piratic Flycatchers were calling loudly in the surrounding trees, while the flatbills repeated incessantly their sharp, thin whistles, unseen in the crown of a guava tree. Piratic Flycatchers sometimes breed in nests of the Sulphury Flatbill and evidently they were trying to capture this one; but while I watched they did nothing but call. I continued on my way into the forest and when I returned an hour later a young flatbill, nearly feathered but still unable to fly, lay on the ground below the nest, doubtless having been pulled from it by a Pirate. It seemed uninjured and I replaced it in the nest beside the single nestling that I found still within.

I then sat down to watch from a good distance. During the next 40 minutes the Pirates repeatedly came into the neighboring trees, calling vociferously. Sometimes they approached quite near the flatbills' nest, but they did not try to enter it. The flatbills made no effort to drive away the somewhat larger Pirates but confined themselves to vocal protests from a distance. While their persecutors were in sight they did not feed their nestlings. But after the intruders withdrew they resumed feeding and in the next hour they brought food 13 times.

Next morning, from 07:00 to 09:00, the pair of Piratic Flycatchers spent much time in the trees near the Sulphury Flatbills' nest, incessantly repeating their breezy *pee-e-e-e* and *pee-de-de-de*. From time to time they approached the hanging nest, revealing a little of the concealed yellow of their crown patches. They seemed timid about entering. After considerable hesitation and many tentative advances one of them would cling, back downward, to the end of the spout or even enter it. This was done repeatedly during the two hours that I watched, but I could not learn whether one or both members of the piratic pair went into the nest. Again the flatbills failed even to try to drive away the intruders. Yet at 09:00 both nestlings were safe in their pensile nursery. When I returned later in the morning, however, only one young flatbill remained within. The other lay on the ground below, already dead and covered with fire ants (*Solenopsis*). By the middle of the afternoon the nest was empty: the second nestling had quite vanished.

After destroying the nestlings the Piratic Flycatchers never used this nest.

Problem of Oropéndolas' Nests

While studying the Chestnut-headed, or Wagler's, Oropéndola on Barro Colorado Island, Chapman (1928:154-161; 1929:111-121) watched

Piratic Flycatchers harass the far larger icterids and gain possession of some of their nests, only to abandon them before any young were hatched. He concluded his detailed account of his observations with the admission that "after four seasons' observations I have, therefore, still to discover how Legatus perpetuates its kind." Doubtless part of the difficulty was that Chapman was in the habit of spending the dry season on the island and usually left about 1 April or earlier, while Piratic Flycatchers lay their eggs principally in April and May, at least in the Valley of El General in Costa Rica, which is at the same latitude as the Panama Canal Zone. Nevertheless, they sometimes lay their eggs earlier if suitable nests are available. Nests are available long before April in colonies of oropéndolas, which are active early in the year.

From 14 March to 24 June 1964 I resided near Cañas Gordas at an altitude of about 3,800 feet on the Pacific slope of southern Costa Rica. The house that I occupied looked over a pasture with a number of tall, scattered trees in which Chestnut-headed Oropéndolas built their long, pyriform pouches. During the whole of this interval Piratic Flycatchers made themselves conspicuous among the trees where the oropéndolas nested. Although I made no special study of the flycatchers, I looked for them to feed nestlings, which is usually a conspicuous activity. On 15 April a Piratic Flycatcher was carrying pieces of leaf into a recently completed oropéndola's pouch which hung about 80 feet up in a tall, clean-boled tree. It was the only oropéndolas's nest in this tree but 50 yards away was a group of about 16 nests. The flycatchers pursued the oropéndolas as they flew past. As late as 6 June I saw a female flycatcher enter this nest and on 22 June a pair showed interest in it. Yet, as far as I could learn, they neither incubated nor fed nestlings in this nest.

A tall, slender tree still closer to the house held 15 oropéndola nests when we arrived in mid-March and a pair of Piratic Flycatchers was harassing the birds who built them. By mid-May the oropéndolas had deserted the colony, which now contained only 11 nests, the others having fallen. A pair of Piratic Flycatchers, however, continued to call much and loudly among the empty nests. They showed their proprietary interest unmistakably late in the afternoon of 31 May when a Blue-throated Toucanet entered the nest tree to search for eggs or young to eat. When the big-billed bird clung to the side of one of the long pouches, they darted again and again past its head. Each time that a flycatcher shot by the toucanet snapped its great mandibles at its assailant. Usually the flycatchers failed to touch the toucanet but once they knocked out some feathers, which floated slowly downward. Undaunted, the green bird continued to investigate the pouches, clinging to their sides sometimes with its body upright and sometimes with its head downward. Once it entered a nest but apparently did not descend to the bottom. All this while the flycatchers continued to dart vociferously past the intruder's head. Finally, having found nothing edible, neither egg nor nestling, the would-be pillager flew back to the forest leaving the flycatchers calling loudly among the swinging nests.

On the following day, 1 June, I twice saw the female Pirate carry a bit of material into one of these nests. By 5 June she seemed to be incubating in this nest and observations on 8 June strengthened my conviction that this was true. On 22 June, however, I could find no indication that the birds were incubating or feeding nestlings in any of these pouches, although the male still called much in or near the nest tree. Two days later he clung to the nest in which incubation had seemed to be in progress early in the month. A lone male oropéndola paused among the empty nests to pour forth his liquid gurgle and as he continued on his way the little Pirate pursued him.

Despite all the interest that the Piratic Flycatchers showed in the oropéndolas' nests over a period of three months in the height of their breeding season, and occasional indications that they were incubating in some of these nests, the only nestlings that I noticed in this interval were in a neighboring nest built by either the Gray-capped or the Vermilion-crowned Flycatcher. In other parts of Costa Rica, too, I have looked in vain for Piratic Flycatchers to carry food into an oropéndola's pouch.

Although there is no lack of information as to how *Legatus* perpetuates its kind, whether it ever does so in oropéndolas' nests is still as uncertain as it was when Chapman concluded his study 40 years ago. There is no doubt that these big, commodious pouches, hanging in conspicuous clusters and not too difficult to wrest from the much larger birds who weave them, strongly attract the Piratic Flycatcher, yet it is questionable whether they are suitable for its reproduction. They are lined with a loose litter of leaf fragments and the like—material similar to that which *Legatus* itself carries in much smaller amounts into the nests it captures. I surmise that the trouble is that the flycatcher's small eggs tend to become separated from each other and from the incubating female, perhaps even lost, amid the shifting materials in the bottom of a basket intended for the use of a very much bigger bird. This is probably the reason why incubation is never, so far as I know, successfully concluded in oropéndolas' nests. In the smaller swinging pouches of such icterids as the Scarlet-rumped Cacique and the Yellow Oriole, miniatures of those of oropéndolas, the Piratic Flycatcher does indeed, as I have myself seen, succeed in hatching its eggs. But these nests are lined with soft down or with grass, on which the eggs rest, rather than with loose leaves into which they would undoubtedly sink.

Then, too, it seems that the abundance of available nests in an oropéndolas' colony diverts the flycatchers' attention from the one on which they should concentrate to rear their brood. In short, the oropéndolas' great pouches act as a "supernormal stimulus" to certain

Piratic Flycatchers, attracting them with irresistible power yet frustrating their efforts to perpetuate their kind.

SUMMARY

The Piratic Flycatcher breeds in covered nests that it captures from other birds by harassing them and finally throwing out their eggs. To the nine host species already recorded, two are now added: the Scarlet-rumped Cacique and the Rufous-fronted Thornbird. A pair of flycatchers and a pair of thornbirds simultaneously attended nestlings in different compartments of a three-chambered nest of interlaced sticks built by the thornbirds.

In addition to removing eggs, as earlier reported, Piratic Flycatchers may throw nestlings from the nest of other small birds. After pulling two nearly feathered nestlings from a retort-shaped nest of Sulphury Flatbills, the Pirates failed to occupy this structure.

Although we have numerous observations of Piratic Flycatchers entering, and apparently incubating in, the great pensile pouches woven by oropéndolas of several species, they seem never to succeed in rearing nestlings in these pouches. Apparently the female flycatcher cannot keep in contact with her eggs lying in a loose litter of leaves in a basket woven by a much bigger bird. The oropéndolas' pouches seem to act as a "supernormal stimulus," frustrating the flycatchers' efforts to reproduce.

WHITE-RINGED FLYCATCHER

Coryphotriccus albovittatus

About six inches long, the White-ringed Flycatcher is named for the band of pure white which, passing across the forehead, above the eyes, and around the nape, completely encircles the dark brown crown, in the midst of which is a concealed patch of bright yellow. The upper parts of the body are grayish olive-green. The tail is grayish brown and the wings are dusky, with paler edgings. The sides of the head are dark sooty brown; the chin and throat are white; and the remaining under parts are bright yellow. The short bill, broad at the base and tapering to a point, is black. The eyes and feet are dark. The sexes are alike.

The color pattern of this middle-sized flycatcher is confusingly similar to that of a number of other species in three different genera, including the Great Kiskadee and Boat-billed Flycatcher, which are much bigger, and the Vermilion-crowned, Rusty-margined, and Lictor Flycatchers, which are of about the same size. Although minor differences in coloration serve to distinguish all of these birds, one familiar with them depends chiefly on their distinctive voices for recognition.

The similar appearance of a number of birds of different genera, several of which are sympatric over wide areas and should, according to evolutionary theory, exhibit character displacement (divergence in conspicuous features or behavior in order to facilitate recognition or minimize competition in the areas of sympatry) provides food for thought. What peculiar advantage does this particular color pattern confer on the flycatchers that wear it, to make it so widespread and resistant to alteration?

The White-ringed Flycatcher ranges from Costa Rica, where it is known only from the lowlands and foothills on the Caribbean side, southward through Panama and western Colombia to Ecuador. It is sometimes considered to be conspecific with *Coryphotriccus* (or *Conopias) parvus*, a yellow-throated form widespread in northern South America east of the Andes.

The only place where I have found White-ringed Flycatchers was La Selva, where they were moderately abundant. Usually I saw them wandering through the tops of the tall trees that stood in the cacao plantations, the pastures, or along the river banks. Sometimes I heard them far above me in the forest where it was scarcely possible to glimpse them through the intervening masses of foliage. Often they travelled in a group of two, three, or four, all of which appeared equally mature. After delaying a while in the top of one tree they would move on to another; they seemed to travel far in the course of a day. They invariably remained high in the trees.

The White-ringed Flycatchers' method of foraging interested me greatly for I had never seen any other flycatcher, or indeed any bird of whatever kind, hunt insects just as they did. Perching conspicuously on the ends of high, exposed, leafy twigs, often the topmost shoots of towering trees, they scanned the foliage around and below them and when they detected an insect, caterpillar, or spider on a leaf or branch, they flew out, or more frequently down, to pluck it off. Often they shifted from one high twig to another in the same or a neighboring tree, to bring fresh foliage into view. Larger victims were beaten against a perch before being swallowed. These flycatchers scarcely ever darted out into the air to catch flying insects, in the manner so frequent among the large and middle-sized members of this family. They varied their diet with a little fruit. Once I watched them gathering the small white berries of a scandent *Tournefortia* that had climbed high into a tree. The birds plucked the fruits without alighting, much as they snatched insects from foliage.

These flycatchers exploit an ecological "niche" or manner of feeding peculiarly their own. A number of flycatchers scan the ground from a low perch and fly down to pick insects and other small creatures from the bare soil or low herbage. Others, especially the smaller kinds, flit among the foliage, often flying up to pluck creeping things from the *under* side of a leaf. Wood warblers, vireos, cotingas, and many other insectivorous birds hunt largely *inside* the crowns of trees. But White-ringed Flycatchers scan the crowns from above, where they look over the *upper* surfaces of the leaves and must often detect insects which other birds miss. Their habit of resting on high, exposed perches would seem to make them especially vulnerable to hawks; but in five months at La Selva I did not see a hawk attack any kind of bird. Indeed, everywhere in the wooded regions of the tropics I have found adult birds remarkably immune from the ravages of diurnal birds of prey.

The call of the White-ringed Flycatcher is a high-pitched, whining or squeaky note, followed by a little churr or rattle in much the same key—a somewhat woodpecker-like utterance. Another vocalization is an almost-clear trill, slowing down until it terminates with several disjoined notes.

Although, to my great regret, I learned nothing about the nesting of this flycatcher, I thought that its unique manner of foraging should be described in this book. In Surinam Haverschmidt (1957) found a single nest of the closely related (perhaps conspecific) *Coryphotriccus parvus*. It was in a hole near the top of a tall dead tree standing above second growth on sandy soil. Neatly constructed of dry grasses, it filled the bottom of the hole. On it rested two eggs "of a cream color, and covered all over with streaks and blotches of chocolate brown, which formed an unbroken, solid, broad ring around the large end."

Summary

In pairs or family groups of three or four, White-ringed Flycatchers wander through the tops of tall, scattered trees or over the roof of high rainforest.

Their method of foraging is rare, if not unique, among flycatchers. Perching conspicuously on the ends of high, exposed, leafy twigs, they scan the foliage around and below them, then fly out or down to seize the insects or spiders they detect, often plucking them from the upper sides of the leaves. They vary their diet with a little fruit.

Their notes are described. Their nest is unknown.

BLACK-HEADED TODY-FLYCATCHER

Todirostrum nigriceps

One of the smallest of passerine birds, the winsome little Black-headed Tody-Flycatcher is only three inches long. In both sexes the head is uniform black. The upper parts of the body are yellowish olive-green. The rectrices and remiges are dull black with yellowish olive-green margins and the wing coverts are glossy black tipped and edged with yellow. The pure white chin and throat contrast with the canary yellow of the remaining under parts. The rather long, broad and flat bill is black; the eyes are brown; the legs and toes are dark.

This tody-flycatcher ranges from Costa Rica to Ecuador and western Venezuela. In Costa Rica it appears to be confined to the Caribbean slope below about 2,200 feet but in Venezuela it has been recorded up to 3,150 feet above sea level (Phelps and Phelps, Jr., 1963:213). Unlike the related Black-fronted and Slate-headed Tody-Flycatchers, it inhabits forest as well as areas of lighter vegetation and scattered trees, such as the fringe of trees and shrubbery along river banks and shady plantations or pastures; but it is so seldom seen in the more open parts, where alone it is somewhat easy to detect, that it has acquired the reputation of being a rare bird. Actually, it is more abundant in the crowns of the tall trees of the rainforest, which appear to be its true home. After I had learned to recognize its call I was surprised how numerous this flycatcher was in the treetops at La Selva, although I never saw it except when it ventured beyond the closed woodland.

The Black-headed Tody-Flycatcher catches its insect food largely by means of short darts amid the foliage rather than on long sallies into the air, in the manner of many larger flycatchers. Sometimes it sidles along its perch, as the Black-fronted Tody-Flycatcher more frequently does. Its call or song consists of a low, measured *chip chip chip* . . . similar to, but more resonant than, the ticking note of the Black-fronted Tody-Flycatcher. I heard these *chip*'s singly or in series of two to 15, but usually 10 to 14, delivered about as fast as I could count. While the male of a nesting pair chipped quite frequently, his mate, sitting in the nest, continually repeated a slight, soft trill which the male rarely used. I did not hear these birds trill back and forth to each other as mated Black-fronted Tody-Flycatchers so frequently do with their louder, clearer trills.

Nesting

The only nest of this flycatcher that I have seen was found on 10 May 1968 in a pasture at La Selva, about 20 feet above the ground in a cedar tree (*Cedrela* sp.) standing apart from other trees. The pensile structure was firmly attached beneath a drooping branch where it was well screened by foliage. Only about six inches away was a small vespiary of gray paper, inhabited by middle-sized black wasps with white-tipped wings. The flycatchers' nest was a short, pear-shaped construction with a side entrance protected by a visor-like projection from the wall. Light in color, it was composed of felted fibrous materials and seed down, apparently largely of the balsa tree. Although similar in form to nests of the Black-fronted Tody-Flycatcher, this structure was shorter with less "tail" hanging loosely below the bottom; it was attached to a thicker branch and better concealed than nests of the related species usually are. This nest in the cedar tree was at least 50 yards from the nearest thicket and even farther from the forest.

On the morning when I found the nest the female twice entered bearing a great tuft of tawny seed-down, almost as big as her diminutive self. She sat in the nest for short intervals and appeared to be incubating, or preparing to do so. I could not reach the nest to examine its contents. A single egg of this species from the Santa Marta region of Colombia is described as "regularly ovate, clear white, with a few yellowish brown or rust-colored specks over the larger end" (Todd and Carriker, 1922:383).

Only the female incubated, as in other flycatchers. In watches of an hour or two each on five different days, mostly at various times of the forenoon but once in the usually rainy afternoon, I spent a total of eight hours making records of her movements. Nineteen sessions ranged from three to 19 minutes in length and averaged 8.8 minutes. Eighteen recesses varied from three to 49 minutes and averaged 16.6 minutes. Calculated from these figures, this female incubated with a constancy of only 34.6 per cent, the lowest that I have recorded for any bird watched for a comparable period (see Skutch, 1962b, Table 2). Much of the time her nest was in sunshine and doubtless uncomfortably warm within, but even when it was shaded her sessions were shorter than her absences.

On leaving her nest the female flycatcher usually rose higher into the crown of the tall cedar tree and vanished. Returning she came down from above, flitted to a twig in front of her doorway, and darted in. At the moment of entering she nearly always uttered a low, soft trill; and often while sitting she repeated this sound suggestive of happiness or contentment. The arrival of her mate on a neighboring branch often set off renewed trilling, showing that she somehow sensed his presence even when she did not see or hear him. From my observation post on the ground I could see nothing of the incubating flycatcher; but from time to time these confidential notes, and the vibration of the nest that accompanied their production, assured me that she was still within.

The male flycatcher did not escort the female to the nest by flying close beside her, seeming to race her to the doorway as males of the Black-fronted Tody-Flycatcher, euphonias, and certain other birds with closed nests do. But he was most attentive, spending as much, or more, time perching a foot or two from the nest, both while she was present and during her absences, as she spent inside the nest. While resting there amid the foliage he often repeated series of *chip*'s, loud for so small a bird, or else he preened. When a Great Kiskadee alighted near the nest he flew with clacking bill at this relatively huge flycatcher, who did not tarry. Once one of the black wasps from the nest so near his own made the tody-flycatcher shift his position, but otherwise he had no trouble with his stinging neighbors. From time to time, especially during the female's absences, he went to the doorway of the pensile nest to peer in and once after doing so he trilled in a low voice.

But all the rest of the time that I was at the nest the two sexes used different calls, the female trilling as much as the male chipped.

Our sojourn at La Selva came to an end before the eggs hatched, but I had no doubt that if the female's so inconstant incubation brought their progeny to life, the attentive male would help to feed the nestlings, as the males of other kinds of tody-flycatchers do.

Summary

Although Black-headed Tody-Flycatchers are sometimes seen in shady plantations and secondary growth, their true home appears to be in the crowns of great forest trees where they are difficult to detect. They catch small insects as they flit and dart amid the foliage.

Their notes include a low *chip*, delivered singly or in series of two to 15, and a slight, soft trill. The former was more frequently heard from the male, the latter from the female sitting in her nest.

A nest was found 20 feet up in an isolated tree in a pasture. Composed largely of felted fibrous materials and seed down, it was a pensile, pyriform structure with a side entrance protected by a visor.

Only the female incubated, but so inconstantly that she was in the nest for only 34.6 per cent of eight hours. Her mate spent much time perching nearby and often went to peer through the doorway at the eggs.

Family CINCLIDAE

WHITE-CAPPED DIPPER

Cinclus leucocephalus

The White-capped Dipper is a stout, long-legged, short-tailed bird about six inches in length. In the nominate race, with which this account deals, the whole crown and hindhead are white. The rest of the upper parts, including the wings and tail, and the sides of the head and neck are sooty brown, with a small white patch on the upper back, separated from the white of the head by a narrow band of sooty brown. At the bases of the remiges are small areas of white, visible only when the wings are spread. The under parts, except the sooty sides, flanks, and under tail coverts, are white. The short bill is black; the strong legs and toes are leaden blue. The sexes are similar, but in a pair which I watched the white on one, doubtless the male, was replaced by pale gray on the other, and the white patch on his back was larger.

The White-capped Dipper lives along the mountain torrents of South America from Venezuela through Colombia and Ecuador to Peru and Bolivia. In Venezuela it ranges vertically from about 6,500 to 11,800 feet above sea level (Phelps and Phelps, Jr., 1963:259). In Ecuador, where I found it on both sides of the Andes, it descended to about 4,000 feet along the Río Pastaza. I did not trace the species to its upper limit. Like their relatives in Central America, the dippers along the Pastaza were usually in pairs, at least from July to October. They fed on small creatures that they found on the wet, moss-encrusted, shelving cliffs above the river and on slippery rocks at the very brink of the current. Often they seemed to tempt fate by standing on a point of rock that was covered by rhythmic surges of the wild torrent; but whenever the angry water reared up, threatening to overwhelm them, they leapt vertically upward beyond its reach and flitted lightly to a higher ledge. I did not see these dippers dive or pick food from submerged stones, as I have watched American Dippers do in the somewhat quieter and much purer waters of certain Central American streams. The strength and opacity of the Pastaza's brown flood discouraged such practices.

Nesting

In the third quarter of the year 1939, a pair of White-capped Dippers was established on the Río Pastaza in the vicinity of the Cascada de Inés María, a short distance above the town of Baños. Here the river, laden with silt from the high Andes, rushes through a profound and narrow gorge, whose right side is a lofty, menacing, scarped cliff, at the foot of which the torrent flows, while the opposite side is a long, bare, precipitous slope rising thousands of feet. At a point where the channel is divided in the middle by a massive pier of gray rock, the river plunges over a wall 15 or 20 feet high into a nearly rectangular, cliff-enclosed pool whose waters are perpetually in a state of violent agitation. From this turbulent cauldron the turbid flood, now contracted into a single undivided cascade, escapes by leaping down a somewhat higher wall of rock. Then it spreads over a broader bed as it rushes down, between lofty fractured cliffs, toward the distant Marañon. Such was the scene which the dippers chose for their nesting.

Their nest was situated in a narrow niche in the highly fractured face of the slightly overhanging cliff on the left side of the first, or upper, waterfall. Since this was the bank of the river away from the town where I lodged, in order to approach the nest site I was obliged to cross the gorge by way of the bridge of San Martín, a quarter of a mile above, then descend hundreds of feet into the canyon. I reached the brink of the waterfall only to find the nest quite inaccessible to me. But I came close enough to see that it was a very bulky structure, open above, and composed largely of moss, now dry and yellow, although it had probably been green when the dippers gathered it.

When I found the nest on 29 September it contained two nestlings, already feathered. Seated on a platform of gray volcanic rock high above the swirling water of the pool between the upper and lower falls, I watched the nest across the impassable current. Through my binoculars I could see parts of the young dippers as they stretched forward in their cranny for food, opening their mouths to reveal an orange-yellow interior. Arriving with its small bill laden with a number of tiny creatures plucked from the wet rocks at the water's edge, the parent would alight on the gently inclined foot of the cliff, hop or walk a short distance upward while it surveyed the gorge to make sure that all was safe, then fly up vertically to alight on the edge of the bulky nest. Here the food was delivered with a very rapid movement and in a trice the dipper flew away again. Both parents fed the young but I did not see either of them remove a dropping.

By the morning of 1 October the young dippers had grown too strong and restless to remain quietly in their nest, but they were not yet ready to attempt the flight across the furious torrent beneath them. So one of them tried to leave the sheltering niche where it had hatched by climbing out on the face of the surrounding cliff. But here the vertical rock face offered no footing. Finding that it had attempted an impossible feat, the young bird, who had retained hold of the nest with one foot, managed with great effort to maneuver itself into an upright position and by continued exertion succeeded in regaining the safety of the nest. I saw one, or possibly both, of the nestlings repeat this hazardous escapade several times in the course of the morning.

At 08:30 the following morning I found one of the young dippers resting on the shelving base of the cliff, below the nest. Probably it had tried to climb out on the rock face once too often, lost its hold on the nest, and fluttered down to its present position. To judge by the number of white splashes on the rock around it, the fledgling had passed the night there. The bird climbed over the rock, sometimes slipping, and from time to time ventured closer to the edge of the cataract than seemed safe for one so young and inexperienced. It continually twitched its wings and sometimes appeared to pick things from the rock, but I do not know whether it found anything edible. Presently, seeing its parents foraging on the mossy rocks of the opposite shore, it boldly flew across the swirling water of the pool below the falls, a distance of about 75 feet. When the parents noticed it on the wing they hurried to it and followed it closely on the final few yards of its first considerable flight, just as swallows and other birds escort their young as they fly from the nest. Safely reaching the opposite cliff, the young dipper clung to an irregularity on the steeply inclined rock surface and seemed quite at ease.

As often happens in such cases, the parents gave most of the food to the fledgling in the open and neglected the stay-at-home. The latter tried several times to leave the niche by climbing out on the vertical cliff, as I had seen on the preceding morning, but always failed. The parents made no effort to coax it from the nest and after a while they fed it more frequently. When I left at 10:00 the second young dipper was still in its niche in the cliff, but by the following morning it, too, had flown.

These fledglings closely resembled their parents. The one who left first had a white patch on its back almost as large as its father's, but its crown was as gray as its mother's. I did not see the second fledgling so well.

Summary

White-capped Dippers forage along Andean torrents, usually in pairs, from about 4,000 to 12,000 feet above sea level.

In September a pair nested in a rock niche in the side of the profound gorge through which the Río Pastaza rushes near Baños, Ecuador. Their bulky nest, open above, was composed largely of yellowing moss. Both parents fed the two nestlings with tiny creatures gleaned at the water's edge. Becoming feathered, the young birds tried vainly to leave their nest by climbing over the surrounding cliff. Finally one flew boldly across the boiling torrent, closely followed by its parents.

Family TROGLODYTIDAE

STRIPED-BREASTED WREN

Thryothorus thoracicus

The Striped-breasted Wren is a slender bird about 4½ inches long. In both sexes the upper parts are brown, becoming more rufescent on the rump and upper tail coverts, more grayish on the wings and tail, both of which are broadly barred with black. There is a narrow white stripe above each eye. The cheeks and sides of the neck are boldly streaked with black and white. The throat and breast are white conspicuously streaked with black, the streaks sometimes extending to the abdomen. The sides and flanks are brown and the pale brownish buff under tail coverts have dusky bars. The upper mandible is largely black but the lower is paler. The eyes are red or brown, the legs and toes are black.

This easily recognized wren is found only from northeastern Nicaragua to Panama west of the Canal Zone. In Nicaragua and Costa Rica it is almost wholly confined to the wet Caribbean side, although in northern Costa Rica, where the continental divide is low, it penetrates sparingly into the Pacific watershed. From the Caribbean coastal plain it extends upward to at least 3,500 feet. It frequents the tangled vegetation along the edges of woodland, the more open parts of the forest, and the fringes of trees and bushes along streams. I noticed it more often along watercourses in the Pejivalle Valley of Costa Rica, at an altitude of 2,000 feet and upward, than in the Sarapiquí lowlands. In the latter region the Bay Wren was a common streamside bird, but in the former it was absent. At La Selva the Striped-breasted Wren was thc only abundant bird amid the low cacao trees which, in contrast to the tall trees that shaded the plantations, attracted few birds. At higher elevations it is found among the coffee bushes, especially near the edges of the plantations. It hunts among the dense tangles near the ground, rarely ascending above 15 feet although once I found a pair building a nest 20 feet up. Like other species of *Thryothorus,* this wren is usually found in pairs which appear to persist throughout the year. Its diet seems to consist wholly of invertebrate animals, especially insects and spiders, which it seeks amid the foliage and vine tangles.

VOICE

The Striped-breasted Wren has two distinct styles of singing, so unlike that they seem to belong not merely to different species but to different families or even orders of birds. The first—to take them in the order in which they are usually heard each day—consists of the slow repetition, at regular intervals, of an identical, full, deep, monosyllabic whistle. These notes are usually delivered in series of from 10 to 20, rarely more or less, at a uniform rate of about three in two seconds, or 1½ notes per second. This monotonous performance is heard chiefly at dawn, especially in April and May, less frequently later in the day. Hearing these mellow, evenly spaced whistles emanating from the depth of a thicket in the dim early light, one expects an antbird or perhaps some small nonpasserine, certainly not a wren. In a variant of this song, which I heard only once, the regularly spaced whistles were alternately high and low, giving a beautiful effect.

The second style of singing is more typical of the large genus *Thryothorus* and earns for our wren an outstanding place even among these superb musicians. The songs of this type consist of beautifully modulated notes differing in pitch. Of such closely articulated verses the Striped-breasted Wren has such a large repertoire that the longer I knew the bird the more impressed I was by the variety and beauty of its vocal productions. Once I heard an exquisite song consisting of about ten notes, but this may have been produced by two wrens singing antiphonally with such perfect synchronization that all seemed to issue from a single throat. Most of the songs are shorter than this. Once when the songsters were on opposite sides of me it was easy to perceive that one delivered four notes while the other, doubtless the female, added two somewhat weaker notes. Such antiphonal singing is widespread in *Thryothorus* and certain other wrens; it serves in place of simpler call notes to maintain contact between mates who forage out of sight of each other amid dense vegetation. The situation in wrens provides convincing corroboration of the theory that song is derived from contact calls (Andrew, 1961:549-555). Juvenile Striped-breasted Wrens, still with yellow at the corners of the mouth, sing a sweet rambling song quite different from that of adults.

The Striped-breasted Wren has a scolding note so similar to that of the House Wren that on hearing it I sometimes looked for the latter, which was absent from La Selva.

Nest and Eggs

In the Pejivalle Valley, years ago, I found a number of nests of this wren placed in bushes and vine tangles, or sometimes on the lower branches of trees, usually near a stream at heights of about five to 20 feet above the ground. At La Selva in 1967 a nest was built on top of a palm frond that projected from the shrubby edge of the forest into the clearing behind the house. It was 10 feet above the ground. In the following year, when this small palm was dying, a nest was constructed very close to it on a slender branch projecting into the clearing eight feet above the ground. Another nest was six feet up in a tangle of vines beside a trail in a rather open part of the forest, near its edge.

The nest is a closed structure similar in form to that of the Riverside Wren (Skutch, 1960a:135). Roughly globular, it is usually saddled over a thin horizontal branch with the rounded chamber on one side of the support and on the other side, serving as a counterpoise to the chamber, a spacious antechamber or vestibule, entered through a wide doorway that faces downward and inward. It is important that the chamber and antechamber be of nearly equal bulk and weight, for if they do not balance each other the nest may pivot toward the heavier side and become unserviceable. The nests of the Striped-breasted Wrens are composed of fine fibrous materials, leaf skeletons, and the like with a more or less complete covering of green moss. As I have seen at several nests, both sexes build, taking nearly equal parts. One nest was practically finished in six days but the builders continued to add a little lining for at least two days longer.

Since these nests are used as dormitories as well as for reproduction, only the presence of eggs or young indicates the breeding season. A nest in the Pejivalle Valley had a newly laid egg on 11 April 1941, but the following day it was broken and no more were laid. In the nest we chiefly studied at La Selva, two eggs were laid about 26 April 1967. Carriker (1910:759) found a set of three eggs, far advanced in incubation, in the Caribbean lowlands of Costa Rica on 9 May 1905. The eggs of the set laid in 1967 were immaculate white, slightly glossy, and measured 21.0 by 14.0 and 20.4 by 14.0 mm. Carriker gave the measurements of one egg as 19.5 by 14 mm.

Incubation

While the nest built behind the house at La Selva in 1968 was still far from finished, one member of the pair, doubtless the female, began to pass the night in it. She slept with her feet grasping the supporting twig and much of her under parts exposed. The antechamber of this nest was made too heavy, with the result that the whole structure rotated forward on the slender supporting twig as an axis. I fastened it in a more normal position with pins, but it was abandoned.

In the preceding year the female wren slept alone in the nest on a palm frond for about a week before she began to lay. I watched this wren incubate through the morning of 9 May, an undertaking which demanded close and unwavering attention for hours together, for she was visible for only a second or two as, coming or going, she flitted rapidly between her nest and the dense vegetation at the forest's edge a foot away. Her first outing of the day began at 05:56. She was absent for 49 minutes, returned and sat for 90 minutes, took a recess of 39 minutes, incubated for 109 minutes, then left for 57 minutes, returning at 11:40 when I suspended observation. So slow was her rhythm that she filled the morning with three long recesses and two longer sessions. I could not distinguish the sexes but from the fact that there was no changeover, and from the character of the record, I concluded that only the female incubated, as in other wrens. The male did not visit the nest, but once as his mate returned he escorted her to a point near it in the thicket. Except for some singing at dawn and throaty notes as the female approached the nest, the wrens were nearly silent on this day.

Nestlings

On 14 May the eggs hatched. The two nestlings were pink and blind, and devoid of natal down. The inside of their mouths was yellow, as is usual in wrens. At the age of eight days these nestlings, in pinfeathers and still without down, were heavily infested with tórsalos, dipterous larvae that formed large swellings beneath their skin. At ten days their plumage began to expand.

On 26 May when the two nestlings were 12 days old and nearly feathered, we watched throughout the morning, which was cloudy with intermittent light showers. Both parents fed them, bringing food 57 times in the six hours following the day's first meal at 05:45. In the first half-hour of activity the young received nine meals but in the second half-hour they were fed only once. From 06:45 to 07:45 14 meals were brought. Although the parents often foraged on the opposite side of the clearing, they never approached the nest by flying directly across the open lawn but always came by a roundabout route along the forest's edge, which kept them well hidden. They exposed themselves so little in passing between the dense foliage at the margin of the forest and their projecting nest that we could not always see whether they brought anything, so we counted every visit as a feeding, as without much doubt it was. As far as we saw, they brought only invertebrates, including long, slender caterpillars, moths, forest roaches, other insects, and spiders. The nestlings called with little *tsip*'s, especially from about 05:30 until they received their first meals. The parents sang little and did not stay to brood on this wet morning.

When I inspected the nest both parents approached with food. Hopping around in the tangled growth close behind me they sang beautifully, despite their full bills. They never tried to lure me away with distraction displays, which are rarely witnessed in the wren family.

These nestlings were last brooded during the night of 24-25 May when their parent covered them sitting far forward, with her striped breast filling the doorway. During the next five nights they were alone in the nest. One left before 06:00 on 30 May and by 09:30 of the same morning the other had departed, both at the age of 16 days. Still stubby-tailed, they could fly a few yards. One somehow made its way across the lawn to the side of the house farthest from the nest, but I easily caught it and placed it in the thicket by the nest. In the rainy evening neither a parent nor the fledglings came to sleep in the nest, as wrens of certain species do. Whether they entered some other dormitory I could not tell. On the following day I found a parent with a fledgling in the bushy undergrowth of the forest at the edge of the bluff above the river, 50 yards from the nest. The young wren flitted and hopped with agility far up into a tangle of vines.

Summary

In the humid Tropical Zone of Nicaragua and Costa Rica, up to an altitude of at least 3,500 feet, the Striped-breasted Wren lurks amid low, tangled vegetation in the more open parts of the forest, at its margins, and along streams. It also frequents plantations of cacao and coffee. It is usually found in pairs, which apparently persist throughout the year. Its food seems to consist wholly of insects, spiders, and the like which it seeks amid foliage and vine tangles.

This wren has two contrasting song types. The first consists of the slow repetition, at regular intervals, of the same full, deep, monosyllabic whistle—a most unwrenlike performance which is heard chiefly at dawn. Songs of the second type are more typical of wrens and especially of the genus *Thryothorus*. The Striped-breasted Wrens have a large repertoire of these beautiful verses which are often sung antiphonally by a mated pair, the two sometimes articulating their notes so perfectly that one seems to hear a single superb musician.

The nest is a globular structure that is usually saddled over a thin, horizontal branch, the closed chamber on one side of the support balanced by the spacious antechamber or vestibule on the other side. The wide doorway faces downward and inward. These nests are usually situated from five to 20 feet above the ground. They are built by both sexes in six or eight days.

In three nests eggs were laid in April. One set contained three eggs, one two eggs, and the third was not completed. The eggs are immaculate white.

Before she lays the female sleeps in the nest. She alone incubates, taking very long sessions, up to nearly two hours, separated by long recesses. The incubation period is unknown.

The nestlings are devoid of natal down and the interior of their mouths is yellow. They may become heavily infested with dipterous larvae that form great swellings beneath their skin. They are brooded by their mother and fed by both parents. Two 12-day-old nestlings were fed 57 times in six hours of the forenoon. As far as could be seen, their meals consisted wholly of invertebrates, including caterpillars, moths, forest roaches, other insects, and spiders.

The nestlings were alone on their last five nights in the nest. They left when they were 16 days old and could fly a little. Neither they nor their parent returned to sleep in the nest.

Family COEREBIDAE

SHINING HONEYCREEPER

Cyanerpes lucidus

The male Shining Honeycreeper, about four inches long, is attired with simple elegance in plain dull ultramarine, with black wings, tail, and thighs, and a black patch covering his lores, chin, and throat. He has dark eyes and his long, slender, downcurved bill is black but his legs and toes are intensely yellow. The female has dull grass green upper plumage, tinged with blue on her head. She has a buffy throat and her whitish central underparts are conspicuously streaked with blue on the chest. Her sides and flanks are grayish green. Her bill is black, her eyes dark, and her legs and toes yellow or greenish yellow.

Although the male of the congeneric Blue or Red-legged Honeycreeper goes into eclipse after the breeding season, in June or July in Costa Rica, in many years of observation I have noticed no similar change in the adult male Shining Honeycreeper, who wears the same colors throughout the year. Such seasonal changes in plumage are not generic or even specific characters. The male Avadavat is unique among the tropical estrildines in going into eclipse (Goodwin, 1962) and, according to Stresemann (1963) in *Nectarinia takazzae* and certain other sunbirds, some populations have an eclipse plumage but others do not.

The Shining Honeycreeper ranges from the Mexican state of Chiapas to northwestern Colombia. A bird of humid forests, it is confined to the Caribbean side of the cordillera in northern Middle America, but in Costa Rica and Panama it also occurs on the Pacific slope. From the lowlands it extends into the foothills and lower mountains. In the Chiriquí highlands of extreme western Panama it is reported by Eisenmann and Loftin (1967a) to be rare above 5,000 feet. It has been recorded in the vicinity of San José, Costa Rica, around 4,000 feet but I have failed to see it in this densely populated, intensively cultivated district. Russell (1964) found the Shining Honeycreeper to be the most common bird in May on the very narrow ridge at the crest of the Cockscomb Mountains in British Honduras, which is wooded with low trees densely covered with bryophytes and ferns. Here it was found from 3,400 feet down the slopes to 1,300 feet, although it was rare below 2,200 feet.

This honeycreeper inhabits the upper levels of damp, epiphyte-burdened forests and neighboring clearings with scattered trees. Less gregarious than the Blue Honeycreeper, it is usually seen singly, in pairs, or in family groups, rarely in larger parties. It is a bird of spotty, unpredictable occurrence. According to Monroe (1968), in Honduras it is known from only two localities: the Omoa region, where it was abundant in 1855 and 1856, and La Ceiba, where it has been found more recently. I had already spent over three years in the Valley of El General in southern Costa Rica before I noticed a Shining Honeycreeper on our farm in October 1942. Increasing here over the years, as surrounding forests have shrunken, it is now an abundant permanent resident. The only other localities where I have found this honeycreeper are in the Sarapiquí region of northern Costa Rica, at the Echandi Lakes at about 2,500 feet, and at finca "La Selva" beside the Río Puerto Viejo in the lowlands. Here, according to Slud (1960: 109) it is a resident, uncommon from January to April and abundant from May to September.

Food

The diet of the Shining Honeycreeper includes a variety of fruits and arillate seeds, nectar, and small invertebrates. The chief source of the latter seems to be very thin, dangling vines, dead or at least nearly devoid of leaves, over which these small birds search methodically, hanging in all attitudes by their prominent yellow feet while they pluck from the bark creatures that are mostly too small to be detected by the watcher on the ground. Thin, exposed, leafless twigs of trees are sometimes diligently explored in similar fashion. In its manner of finding its insect food the Shining Honeycreeper differs from its congener of similar size, the Blue Honeycreeper, which more often hunts in the foliage at the tips of slender branches, warblerlike. Both of these honeycreepers frequently bend down to scrutinize the undersides of somewhat thicker branches along which they hop, in the manner so widespread among the brilliant little tanagers of the genus *Tangara* and certain other members of the family. They also probe into small, curled, dead leaves.

To procure nectar, and perhaps small insects that lurk in flowers, the Shining Honeycreeper does not, like hummingbirds, poise on wing before the blossoms. Instead it perches behind and above them, leaning forward to insert its sharp bill. At La Selva, where the rubiaceous shrub *Hamelia patens* was abundant in the cacao plantations, sometimes becoming a small tree 30 feet high with a trunk six inches thick, I often watched these honeycreepers probing the narrow, orange-red floral tubes in this fashion.

Like the other four species of honeycreepers resident at Los Cusingos, the Shining is fond of bananas. The first that I noticed at our feeding shelf was a male who came to eat this fruit on 1 November 1944, a dismal day of continuous rain following a long series of wet, gloomy days in October. In the ensuing months this or a similar bird appeared on the board at long intervals. Finally, on 24 March of the following year he arrived with a mate, who for several minutes stood beside him on the board, picking pieces from a ripe plantain. For a number of years after this, Shining Honeycreepers visited the feeder chiefly during the heavy rains following the autumnal equinox and thence into the following dry season, but they were rarely seen in the first half of the wet season when most birds were nesting. Not until 1960 did I obtain conclusive evidence of breeding in the vicinity. Finally, a decade later, a pair nested in sight of the feeder.

Among the favorite foods of honeycreepers of a number of kinds are arils, the fleshy or oily outer coats of the seeds of many trees, shrubs, and vines, especially in the tropics. Shining Honeycreepers join related species in eating the white arils around the shiny black seeds of *Dipterodendron elegans,* a tree of the soapberry family. Among the 24 kinds of birds, ranging in size from manakins to toucans and oropéndolas, that I watched eat the seeds of the euphorbiaceous tree *Alchornea costaricensis* were Turquoise Dacnises, Green Honeycreepers, and Shining Honeycreepers. The last-mentioned were almost constantly present in the tree, swallowing whole the small, hard, indigestible seeds covered with red, soft, digestible arils.

Most eagerly sought by honeycreepers are the seeds of *Clusia,* a large genus of mostly epiphytic trees and shrubs that abound in Costa Rican forests. When ripe the fruits, ranging in size from that of an olive to that of an apple, open by from four to 12 valves (according to the species) that spread outward like the pointed petals of a flower, exposing the contents of as many cells packed with small seeds embedded in orange or bright red arils. Almost tasteless or slightly bitter to the human tongue, this soft pulp is so compellingly attractive to a variety of birds that they neglect ripe bananas spread on the board for them to hop around waiting for the Clusia pods to open. With their long, slender bills honeycreepers can extract seeds through the first narrow clefts of the expanding pods, while they are still inaccessible to thick-billed tanagers; but the strong-beaked woodpeckers have first choice.

Some years ago, while a tree of the large-fruited *Clusia rosea* growing on a calabash tree in front of our house was opening its pods like stars with from nine to a dozen rays, it became evident that a dominance hierarchy had developed among the birds that eagerly awaited their dehiscence. At the top were the Golden-naped Woodpeckers, who could displace almost equally big Red-crowned Woodpeckers from the opening pods, and of course all smaller birds. The Red-crowned Woodpeckers, in turn, took precedence over all the honeycreepers. Among the latter, Green Honeycreepers were clearly dominant over the smaller Shining and Blue Honeycreepers. Although the last two differ little in size, Shining Honeycreepers could displace Blue Honeycreepers from a fruit, but the latter could not drive the former away. A Shining Honeycreeper might, indeed, give way to a Blue Honeycreeper that suddenly bore down upon it, but it soon returned to chase away the intruder. Thus the order of precedence at this tree was Golden-naped Woodpecker, Red-crowned Woodpecker, Green Honeycreeper, Shining Honeycreeper, Blue Honeycreeper. The last-mentioned were the most numerous but least aggressive of all the visitors to the Clusia fruits. Although tanagers, saltators, manakins, and other birds with short, thick bills eagerly devour the seeds of Clusias with more numerous, smaller pods, they hardly had a chance at this tree crowded with birds able to extract the coveted arils through the first narrow fissures between the expanding valves.

Shining Honeycreepers threaten their adversaries with spread wings, although they have no concealed bright color on their wing feathers to display as Blue Honeycreepers have. Sometimes one Shining Honeycreeper drives another of its own kind from a Clusia pod, but I have never seen them fight among themselves. Neither have I seen any other dacnid honeycreeper fight with a member of its own species—the seizures so frequent among Green Honeycreepers (Skutch, 1962a: 92-93) seem to be an expression of excessive, and sometimes misdirected, sexual ardor. But near the Clusia tree on a May morning, I watched a real fight between a female Blue Honeycreeper and a female Shining Honeycreeper. The two clutched and fell earthward but separated before striking the ground. Then one chased the other around until they returned to the Clusia tree. Here the Blue Honeycreeper postured, repeating her nasal *chaa,* as one frequently sees in nonviolent contests between two individuals of this species, always of the same sex. Her opponent, the Shining Honeycreeper, postured silently in front of her. A male of each kind looked on without participating in the affray. Soon all four flew to a neighboring tree and, as far as I saw, quarreled no more.

Although at the Clusia tree Shining Honeycreepers clearly dominated Blue Honeycreepers, the latter are not always submissive. One morning in June I watched a female Blue Honeycreeper threaten a male Shining Honeycreeper with nasal *chaa*'s. Facing his adversary, he leaned backward, displaying his bright yellow legs prominently to her. Then he fled with the female in pursuit.

Voice

The Shining Honeycreeper is a very quiet bird. The most sustained vocal performance that I have heard was given by a male who, late on a drizzly September afternoon, perched on a thin dead twig at the top of a tall tree in second-growth woods. For ¼ hour he continued to repeat the same slight note at the rate of about once a second. It was a performance of the utmost monotony, quite devoid of melody. For a while a female rested in the treetop several yards from him, but she flew away in the midst of his declamation, which probably should be considered his song. It reminded me of the tuneless dawn song of the Blue Honeycreeper but was even simpler, consisting of only one kind of note instead of two. The hour of the day and season of the year hardly seemed appropriate for an outburst of song. If the Shining Honeycreeper is capable of a more tuneful utterance, I have never heard it.

In the course of building, incubation, and feeding the nestlings both members of a pair were consistently silent. I first heard their voices when I stood below their nest at the time when the nestlings were about to leave. Then they flew down into a shrub about two yards from me and both rapidly repeated a low, hard note with a metallic quality, a *click* or *tick*, which they continued as long as I remained there. Russell (1964) heard from the Shining Honeycreepers a high, cricketlike *zee zee* repeated by many voices.

Nest Building

I first had evidence of the breeding of the Shining Honeycreeper in the Valley of El General on 30 September 1960, when a female fed a fledgling near our house. A decade passed before I found the first nest. On 13 June 1970 I noticed a female carrying material into an out-jutting, drooping branch of a tall timber bamboo growing beside the forest in front of my study window. She had started a nest about 20 feet up, amid thin terminal twigs and clustered foliage that hid it well. From 07:00 to 08:00 that morning she came to the nest six times, and in the following hour seven times. On each visit she brought a thin piece of material, probably a fibrous root of an epiphyte. In earlier years I had watched Shining Honeycreepers gather such material from trees, but they always carried it beyond sight. After depositing her latest contribution in the nest the building female spent from 45 seconds to 2½ minutes shaping the structure.

Each time the female came to the nest her mate followed and perched from a few inches to a few feet away while she worked there, then accompanied her closely as she flew away. Several times while watching the female work he picked up, from amid the branches, a small bamboo leaf which usually he promptly dropped, although once he carried it away, following the female. The female was never seen to bring such leaves and the finished nest contained none. Twice the male came to the nest while his mate was absent, as though expecting to find her there, then promptly left. Once in a neighboring poró tree this male and two others engaged in a chase.

On the following morning, in the two hours from 06:30 to 08:30, the female brought material to the nest only seven times, on each visit working at it from 40 seconds to 2½ minutes. The male escorted her on every visit except one. On this day the very slight structure seemed to be finished; I saw no more building on the following days.

The completed nest was a shallow cup attached by its rim to thin horizontal twigs, in form much like a manakin's nest. The fabric, composed of slender, dark strands, was so thin that much light passed through it; it was hardly more than a fine-meshed net. Cobweb fastened it to the supporting twigs. A foot above it was a small wasps' nest. In its shallowly cupped form this nest resembled those of the Blue Honeycreeper, the Green Honeycreeper, the Turquoise Dacnis, and the Scarlet-thighed Dacnis, but it was more slightly built and more transparent. I believe that we now have enough information to discredit all earlier attributions of deep, pouch-like nests to dacnid honeycreepers (Skutch, 1962a).

Eggs and Incubation

At 06:25 on 17 June I found the female sitting in her slight structure. Twenty minutes later she had gone leaving an egg that had not been present on the preceding day. She incubated the single egg little on the day she laid it, but that night she slept over it. Before 07:21 next morning she laid another egg, completing her set of two. These eggs could be clearly distinguished through the bottom of the nest, which was fortunate, as it was inaccessible and too closely surrounded by twigs to raise a mirror above it. The eggs looked rather dark in color.

On 23 June we watched the nest from dawn to dusk. The morning was clear but the afternoon cloudy, as usual at this season, with a shower that lasted an hour and 20 minutes. Only the female incubated. During the first six hours of the day she took eight sessions, ranging from 14 to 42 minutes, and nine recesses, lasting from seven to 23 minutes. Then she sat continuously for 390 minutes, from 11:23 to 17:53. After an absence of 13 minutes she returned in the failing light at 18:06 and settled on the nest for the night. During the whole day she took nine sessions, ranging from 14 to 390 minutes and averaging 69 minutes. Her ten recesses varied from seven to 23 minutes in length and averaged 14.3 minutes. She kept her eggs covered for 81.3 per cent of an active day of 12 hours and 44 minutes.

On 28 June we again watched this nest throughout the day. From her first departure in the dim early light until noon, the female took nine sessions ranging from six to 38 minutes, and 10 recesses from seven to 34 minutes long. Then she sat continuously from 12:00 to 16:02, during which a shower fell for 1¼ hours. This long session was followed by a 19-minute outing, returning from which the honeycreeper incubated continuously from 16:21 until nightfall two hours later. Considering the day as a whole, she took 10 sessions ranging from six to 242 minutes and averaging 46.5 minutes. Her 11 recesses ranged from seven to 34 minutes and averaged 18.3 minutes. During an active day of 11 hours and six minutes, she sat with a constancy of 69.8 per cent.

While we made the all-day record of incubation on 23 June, we saw no male Shining Honeycreeper. On 28 June the male twice visited his incubating mate before the middle of the forenoon. When he alighted near the nest the second time he appeared to have food in his bill. As on his first visit, the female flew away when he arrived. After delaying in the same spot for a fraction of a minute, he followed her with whatever he had brought. He tried to drive away other male Shining Honeycreepers who came within about 20 yards of the nest. One instance of this has already been given in the account of building. Three days after the beginning of incubation a second male joined the breeding pair in the top of a mango tree near the nest. The resident male postured and flitted his wings until he displaced the trespasser from his perch. Following him, the resident male again made him move. Soon both flew off together without having touched each other.

One or possibly both eggs hatched on the morning of 30 June and the second had certainly hatched by the following day, after 12 or 13 days of incubation, as in other honeycreepers.

Nestlings

Both parents promptly began to feed the nestlings. I spent the morning of 2 July watching them attend the nest. The female alone brooded for 21 intervals ranging from one to 19 minutes and averaging 6.7 minutes. The nestlings, one or two days old, were left exposed for 21 intervals of four to 27 minutes, averaging 10.9 minutes. Thus they were covered for only 37.9 per cent of the bright, warm morning. In six hours and nine minutes, the male fed the nestlings 17 times and the female 22 times, a total of 39 feedings, or at the rate of 3.2 meals per nestling/hour.

The food was carried inside the parents' mouth or throat where it could not be seen. Each parent always delivered whatever it brought directly to the nestlings. Early in the morning the female, continuing to brood, kept her mate waiting near the nest holding food, once for 13 minutes, once for five minutes, and twice for three minutes, until she left and he could reach the nestlings. After 09:00 the brooding female always left as he approached, even if she had been sitting for only two minutes, and he never had to wait for the nestlings to be exposed. All morning I heard not a sound from the parents.

In the bright morning sunshine of 4 July, after torrential rain on the preceding afternoon, I found the female clinging back downward beneath her nest, picking off objects too small for me to see, probably ants. After finishing with the bottom she went around the sides, engaging in the same activity. After working for about 10 minutes at the nest she proceeded to pick things from the surrounding twigs. I have watched similar de-anting of the nest in a number of other birds.

In the first six hours of 8 July, a morning that began darkly with an hour-long shower but finally became sunny, the nestlings were fed at least 31 times by the male, at least 36 times by the female, and three times while the light was so dim that I could not distinguish the sexes, a total of 70 times from 05:20 to 11:20, or at the rate of 5.8 times per nestling/hour. The most rapid feeding was from 08:00 to 09:00, when 14 meals were brought; the slowest in the preceding hour, when the nestlings were fed nine times. Now the food was carried conspicuously in the parents' long, black bills. It appeared to consist chiefly of insects, some of which were hardly visible; but sometimes the parent came with bright red objects, doubtless Clusia seeds, lined up for the whole length of its bill. A few small, dark bodies delivered to the nestlings were evidently berries. The female brooded only during the gloomy and partly rainy early morning, for seven intervals, ranging from one to 10 minutes and totaling 36 minutes. Again she plucked invisible objects from the outside of her nest, sometimes while hanging back downward beneath it. Both parents removed droppings, usually swallowing them but sometimes carrying them away in their bills.

At noon on 12 July one nestling, already well feathered, was perching on its nest's rim looking out while the other rested inside. Although I had never concealed myself while watching these birds, now for the first time they revealed awareness of my presence and uttered the first notes that I heard from them since I found them building. While I stood below the nest, looking up at it, both parents came and fed the nestlings, after which they flew down and complained with hard *click*'s, as already told.

In the bright sunshine next morning, from 06:30 to 09:30, the male fed the nestlings 21 times and the female 38, a total of 59 feedings in three hours, or at the rate of 9.9 times per nestling/hour. Now red or yellowish pulp, evidently arillate Clusia seeds, formed the bulk of the young honeycreepers' food. Sometimes these seeds were lined up from the parent's gape to beyond the tip of the bill, making it look red and abnormally long. Dark berries and small insects completed the nestlings' diet. They had become restless, preening much, stretching legs and wings, and rapidly flapping the latter while they stood on the nest's rim, sometimes both together. Thrice one flew from the nest to a twig a few inches away, only to turn promptly around and fly back to the nest. These spells of activity alternated with intervals of quiet repose in the nest.

That same afternoon of 13 July a young honeycreeper left the nest, gradually making its way through the close-set bamboo twigs until, by evening, it was out of sight. Next morning the second fledgling rose above the nest from twig to twig until, after several hours, it had vanished. The nestling period was between 13 and 14 days, as in the Blue Honeycreeper and Turquoise Dacnis.

The nesting season of the Shining Honeycreeper continues until late in the year, after most other birds have stopped breeding. Once at the very end of September, and in three different years in October, I have seen parents feed juveniles. On 24 October 1963 a female brought two well-grown young to the feeder. Noisier than their parents, they buzzed harshly while waiting to be fed and vibrated their wings as bits of banana were placed in their mouths. On 5 and 6 October 1965 a male in full nuptial plumage led a single fledgling to the feeding board. On 12 October 1968 I watched a female feed two well-grown young at the forest's edge.

SUMMARY

The Shining Honeycreeper inhabits wet forests and neighboring clearings with scattered trees from lowlands up to, rarely, 5,000 feet or a little higher. Its distribution is spotty and unpredictable.

It searches for insects chiefly on very thin, dangling, dead or at least leafless vines and on slender dead twigs of trees. Like other honeycreepers it is eager for the soft arils that cover the hard seeds of numerous tropical trees, shrubs, and vines, especially for those of *Clusia* spp. It comes to feeding shelves for bananas. It sucks nectar from flowers while perching beside or above them. At crowded food trees it displaces the Blue Honeycreeper but is itself displaced by the Green Honeycreeper.

The Shining Honeycreeper is a very quiet bird. A monotonous series of identical notes, heard only once, may be its "song." Otherwise the only notes heard by me were sharp, complaining *tick*'s delivered by parents anxious about their young.

The single known nest, a very slight, shallow cup, was built by the female attended, but not helped, by her mate. She alone incubated the two eggs, sitting with a constancy of 81 and 70 per cent during two all-day watches. The incubation period was 12 or 13 days.

Both parents fed the nestlings, the female slightly more than the male. At first their food consisted largely of insects but red pulp, doubtless arillate seeds of Clusia, became increasingly prominent in their diet as they grew older. The rate of feeding increased from 3.2 times per nestling/hour when they were a day or two old to 5.8 times at eight days and to 9.9 times at 13 days. The nestling period was 13-14 days.

In the Valley of El General nesting begins in June, if not earlier. In rainy October parents were repeatedly seen feeding well-grown young.

Family ICTERIDAE

SCARLET-RUMPED CACIQUE

Cacicus uropygialis

The male Scarlet-rumped Cacique is a slender bird about 8¾ inches long. His plumage is everywhere deep black except the rump, which is flame-scarlet. His sharp bill is greenish white, his eyes are light blue, and his legs and toes are black. The female is slightly smaller and her rump may be paler, but in many pairs the sexes can hardly be distinguished in the field. When the caciques perch their flaming rumps are nearly or quite covered by their folded wings. In this they differ strikingly from their similarly-attired neighbors, the male Scarlet-rumped Black Tanagers, whose equally brilliant rumps are vividly exposed while they perch.

This slighter relative of the Yellow-rumped Cacique ranges from Nicaragua to Venezuela and Peru. In Costa Rica, where it inhabits the wetter parts of both coasts, it hardly ranges up to 2,000 feet above sea level but in Venezuela it occupies the Subtropical Zone, from 4,300 to 7,500 feet (Phelps and Phelps, Jr., 1963:341). Perhaps this pronounced difference in the life zone it inhabits is another reason, added to certain differences in morphology, for regarding the northern form, *microrhynchus,* as a distinct species, sometimes known as the Small-billed Cacique.

Scarlet-rumped Caciques wander in family groups or larger, straggling flocks through the tall forests, where they usually remain high but occasionally descend low in search of food. Like most high-foraging birds, they are by no means restricted to woodland but frequently enter plantations and open country with scattered trees. Often they associate loosely with such other treetop birds as oropéndolas, Black-faced Grosbeaks, Purple-throated Fruitcrows, and White-fronted Nunbirds. Restlessly active, they seldom tarry long in one place. Their wing beats, especially of the males, are sonorous as in many of the larger icterids.

As the day ends Scarlet-rumped Caciques gather in larger, more compact flocks to seek their roosts. In the autumn months of 1947 I dwelt for a while at Piedras Blancas, in the Pacific lowlands of Costa Rica between the Golfo Dulce and the Río Grande de Térraba, where these birds were abundant. Behind the little settlement forested hills rose abruptly and in front, beyond the railroad, stretched the great banana plantations of the fruit company. Every evening the caciques congregated in the tops of certain tall trees at the edge of the forest on the hill behind the buildings. Then, in small groups, they dived down to the lower ground and skimmed rapidly over the tips of the banana leaves beyond the tracks. So low did they descend that at first I supposed that they entered among the banana plants to roost. But after passing over the plantation for several hundred yards they crossed the Río Esquinas and rose into the vine-draped crowns of some tall trees on the farther bank, where many passed the night. Every morning before sunrise they flew back over the banana plantation to the forested foothills.

To bathe caciques descend from the treetops into woodland streams. Once I watched a female Blue-crowned Woodnymph hummingbird dip repeatedly into a little pool in a forest rill, taking her bath. Then a cacique stood in the middle of the shallow rivulet and bathed, noisily splashing the water over itself with its wings.

Food

Scarlet-rumped Caciques search for insects oriole-fashion among the foliage of forest trees. Where these trees are heavily laden with epiphytes, as in the Caribbean lowlands of Costa Rica, the birds find many small invertebrates by probing among these air plants. They often hang with head or back downward to reach their food. They may draw in a curled leaf and hold it beneath a foot while they probe it for hidden larvae or spiders, and larger insects are similarly held while they are eaten bit by bit. Once I watched a cacique cling sideways to the edge of a hanging, ribbon-like segment of a frond of a young pejibaye palm while it plucked larvae or pupae from an elongated wasps' nest fastened to the lower or inner side of the green strip. Apparently it had just torn the thin carton covering from the vespiary but I did not see this. While this cacique was busy with the wasps' brood its companion rested on the bare ground nearby, appearing to eat earth or grit.

Close by the Pacific Ocean at Punta Dominical I watched a flock of caciques foraging in a roadside tree. On a high branch one of them found a large, white, frothy mass made by a spittle insect that was evidently hiding in its midst. With its sharp bill the cacique pulled out long, viscid strands of spittle and wiped them off against the branch. Repeatedly it held an outdrawn skein beneath a foot. After continuing this for a few minutes the bird flew away with its bill and feet sticky with spittle, leaving the insect safe within its frothy citadel. Evidently the foamlike secretion is an effective means of protecting these cercopids from birds.

Caciques also eat a variety of fruits and were among the 24 kinds of birds that I watched come to a tree of *Alchornea costaricensis* to gather the hard little seeds enclosed in thin, red, digestible arils. In the forest near Piedras Blancas was a tall tree overgrown by a vine of *Marcgravia*. The crowded, slender, dangling branches of the vine bore long-stalked green flowers in close whorls, beneath each of which was a cluster of nectar cups, as is characteristic of the genus. In late October and November the chief visitors to the nectaries were Scarlet-rumped

Caciques and Green Honeycreepers, both of which came continually throughout the day. The caciques flew up singly or three or four together, repeating the clear whistles by which they communicated with their companions foraging in the distance. Sometimes they clung to the nectaries themselves and with their bodies hanging below reached up to immerse their pale bills in the green cups. At other times they grasped the stem above the inflorescense and stretched down their inverted heads into the cups. Whether the chief attraction was the sweet fluid itself or the insects that came to drink it, I could not tell, but doubtless the birds took advantage of both kinds of nutriment. The great height of the tree made it difficult to ascertain whether the caciques assisted in pollinating the *Marcgravia* blossoms, which are classic examples of ornithophilous flowers but in certain species are largely self-pollinated in the bud. At least the caciques' crowns were dusted with pale pollen. In addition to the more numerous Green Honeycreepers, the caciques shared the nectaries with Golden-naped Woodpeckers, wintering Baltimore Orioles, and hummingbirds, chiefly the Blue-crowned Woodnymph.

Voice

The cacique's most frequent call is a loud, bright note, suggestive of urgency and excitement. Often such notes are delivered in pairs or in rapid series. What I take to be its song begins with a note hardly distinguishable from the call note, which rises notably in pitch as it is rapidly repeated, to be followed at times by several similar notes falling in pitch. A variant is a liquid crescendo followed by a falling phrase of the same type. Another song is a sequence of higher, bell-like notes hurriedly poured forth, sometimes accelerated and attenuated almost to a buzz. While courting the male utters a series of shrill notes, higher in pitch than the usual song, sometimes with a whining quality. Once a courting cacique delivered a whistle that reminded me of the meadowlark's song, although it was shorter. The variations in the cacique's vocalizations are many but nearly all are clear and bright, inclining to be piercing rather than mellow, fit expressions of the restless vitality of these trim black and scarlet birds that wander widely through forests and clearings.

As in a number of other icterids, females sing much as the males do but their voices tend to be higher in pitch. Often they sing while sitting in their swinging nests.

Scarlet-rumped Caciques are songful throughout the year. At the height of the rainy season in October and November, when scarcely any birds were nesting, their cheerful whistles rang through the lowland forests near the Golfo Dulce, alleviating the gloom of the darkest, wettest days. They and the Riverside Wrens were almost the only birds that sang in this dreary weather. The voice of this cacique has some of the bright, shining quality of that of the Yellow-rumped Cacique but it is not so mellifluous. A less inventive improviser than his larger relative, he has a less varied and amusing repertoire. I refer to the Yellow-rumped Caciques of the race *Cacicus cela vitellinus,* as I heard them in Panama. Although the Yellow-rumped Caciques of the nominate race that I heard in Venezuela likewise produced a surprising variety of sounds, the only notes that I heard from them, on short acquaintance, were sharp and unmelodious.

Courtship

Even in October and November the male Scarlet-rumped Caciques in the rainforests of Golfo Dulce sometimes courted the females. At times the courtship display is preceded by a pursuit. When the female comes to rest the male alights on a perch facing her. Bowing slightly forward he elevates his tail, spreads and loosely waves his wings, while with widely open bill he emits a flood of shrill, screeching notes less musical than his usual song. At La Selva I once witnessed this courtship display in late April, when young of the year were full grown. The courtship posture is of a type widespread among the Icteridae but the forward bow is much less pronounced than in the Montezuma Oropéndola, the wings and tail are not raised so far.

Nest and Eggs

The nesting season of the Scarlet-rumped Cacique begins much earlier than that of most small birds in the Caribbean lowlands of Costa Rica and Nicaragua. In the latter country Richmond (1893) saw a female building in February. At La Selva in mid-March I watched parents feed juveniles almost as big as themselves, and the nests that I found at this time were already abandoned.

Unlike the colonial-nesting, polygamous Yellow-rumped Cacique, the Scarlet-rumped Cacique nests solitarily and appears to be monogamous. Although once I found two deserted nests hanging close together, they may have been built by the same bird; and all the others that I saw were far from any similar nests. The long pouches usually hang conspicuously from high, exposed branches, sometimes at the very top of a tree 100 feet high but occasionally they are low. The lowest of all that I saw was attached to the drooping tip of a spiny frond of a young pejibaye palm growing in a clearing close to the forest; its bottom was only 11 feet above the ground. Another nest was fastened to a drooping branch of a slender vine, from which it hung free in an open space at a height of about 60 feet, in a cacao plantation near a stream.

Caciques, like many other tropical birds with closed nests, often build close by an occupied vespiary, so that the stinging insects may protect them from climbing animals. Sometimes suitably situated wasps' nests are in short supply and birds compete for them. On a slender, horizontal branch of a laurel tree (*Cordia alliodora*) in a pasture was a small nest of gray paper, inhabited by black wasps of medium size. About two feet from this vespiary a Yellow-margined Flatbill built its pensile, black, retort-shaped nest in April; but some mishap befell it before the young could fly and for a while it hung deserted. Then, on 10 May I found it on the ground and a newly finished cacique's nest hanging just where it had been. The same distance on the opposite side of the vespiary was an unfinished nest of the White-winged Becard; whether the caciques were responsible for its abandonment, I cannot tell. Later a Gray-capped Flycatcher started to build midway between the becard's nest and the vespiary, or about a yard from the cacique's still unoccupied nest, but this construction was also abandoned before completion. At the time when the becard, the cacique, and the flycatcher started to build in the laurel tree it was renewing its foliage and almost leafless.

At the end of April and beginning of May I spent four hours watching a nest under construction. All the building was done by the female cacique, who worked at a most leisurely pace, bringing material only six times in the hour when she was most active. As far as I saw she gathered her long fibers from epiphytes high in trees. On each visit she spent much time in the open sleeve that her nest had become, weaving in new strands at its lower end. While she worked her mate always perched close by singing and sometimes plucking insects from neighboring branches. When she flew off for more material he usually accompanied her out and back, although sometimes he waited near the nest until she returned. This late nest, found at an early stage of construction on 28 April, advanced very slowly, but by 15 May it seemed finished. Another late nest, that besides the vespiary in the laurel tree, was built far more rapidly. On 2 May there was no sign of a cacique's nest there but when next I visited the tree on 10 May the pouch was finished and at least partly lined, the whole having been done in a week or less.

A fallen nest that I examined was a long, pensile pouch, beautifully woven—a smaller replica of a Montezuma Oropéndola's nest. From the point of attachment at the top to the rounded bottom it measured 25 inches, while from the lower end of the elongated orifice in its upper part it was 15 inches deep. Near the bottom it was 5 inches in diameter and from this point it tapered upward to 3 inches in diameter at the lower lip of the aperture. This vertically elongated opening was 10 inches high by 1¾ inches wide. The pouch was composed of stiff, fibrous materials, predominantly brown in color. They included the branched inflorescences of some plant and the wiry roots of epiphytes. These materials had been woven into a thin but strong, open fabric, surprisingly uniform in texture, through which much light passed. Long fibers dangled loosely below the pouch. Inside the bottom was covered with a thick pad of light-colored seed down, on which the eggs rested.

The single nest that could be reached contained two eggs on 24 April

FIG. 15. Nest of Scarlet-rumped Cacique hanging from the tip of a branch of a laurel tree (*Cordia alliodora*). On the same branch are a small vespiary, an unfinished nest of the Gray-capped Flycatcher, and an unfinished nest of the White-winged Becard (nearest trunk). Near Puerto Viejo de Sarapiquí, Costa Rica, May 1968.

1967. These eggs were white marked with a few light brown and blackish spots and some short scrawls on the thicker end, still fewer elsewhere.

Incubation

Only the female incubates. During six hours of the morning of 26 April the cacique whose low nest was attached to a palm frond took six sessions, ranging from 18 to 54 minutes in length an averaging 38 minutes. An equal number of recesses varied from 15 to 26 minutes and averaged 20.3 minutes. This female incubated with a constancy of 65 per cent. Returning to her swinging pouch she fluttered down from above and dived in headfirst, hardly pausing at the narrow doorway. Sometimes while she sat I could dimly discern her through the meshes of the nest fabric. Departing, she flew from the doorway, going usually toward the neighboring river, but sometimes into the still nearer forest. While covering her eggs she often sang, sometimes in a voice hardly to be distinguished from her mate's, sometimes at a higher pitch and more rapidly. Once, just before leaving, she continued to sing loudly for six minutes. At times the male and his sitting partner sang responsively.

On five of the female's seven returns to the nest she was either accompanied by her mate or I heard in the vicinity song that I took to be his. On two returns she seemed to come alone. Five times the female left her nest while her mate was singing nearby and they flew off together, or else she heard a song, evidently his, in the distance and flew toward it. Twice she left when I had no indication that the male was within hearing. Thus the male was attentive to the female; but he did not once go to the nest, and the longest time that he spent in the surrounding trees was about five minutes. Mostly he was neither seen nor heard while she incubated but returned in time to accompany her when she went to seek food.

Young

On 1 May this nest held two nestlings which had possibly hatched on the preceding day. Their pink skin was devoid of down and the interior of their mouths was pale pink, much like the skin on their bodies. The flanges at the corners of the mouth were pale yellow. On the following morning I watched this nest for three hours. On her five returns the female dived into it so swiftly that I could not see whether she brought anything for her nestlings, although undoubtedly she did. She brooded them for intervals of 12 to 16 minutes. The male was much less attentive than on the morning when I watched the female incubate; I saw and heard him only once and he did not approach the nest.

On 3 May, when the nestlings were three or four days old, their pinfeathers were sprouting, their eyes partly open. While I examined them their mother arrived with a white larva or pupa in her bill. Alighting nearby she made no protest and after I withdrew a short distance she carried in the food, then flew away with her mate. Later she took another meal into the nest while I stood only three or four yards away.

While we inspected the nest on 5 May the parents arrived together, the female with a large spider, the male with a tiny object scarcely visible in his bill. She entered while three people stood close by, then flew off. He followed without having delivered what he held.

The male cacique seemed to be slowly working up to the point of feeding the nestlings. Three days later I arrived at dawn expecting to see him do so; but, alas! the nest was lying on the ground, sodden with rain, and the nestlings were dead. A branch falling from a neighboring tree had knocked down the palm frond to which the pouch was attached. I hoped to study the care of the young at the late nests, even if they were inaccessible; but these laboriously woven pouches were occupied, if at all, only after a long interval. By early June there was no indication that incubation was in progress at a high nest that had seemed finished in mid-May. And the nest that had been so rapidly built beside the vespiary at the beginning of the following May hung unattended for the remainder of the month. By May early broods had long been on the wing and these late nests had evidently been built in anticipation of second broods that never materialized. Although I saw no evidence for late broods at La Selva, Arbib and Loetscher (1935) found this cacique breeding in July or August in the Panama Canal Zone.

Well-grown juveniles differ from adults chiefly in having duller red rumps and dark rather than blue eyes. They still accompany their parents and are apparently fed by both of them. In mid-March I watched such a family. One of the parents pecked open the long stipular sheath enclosing the youngest leaves at the apex of a twig of a chumino tree (*Pourouma aspera*). From this green envelope it removed an insect which it passed to the young cacique, who received it with fluttering wings. The juvenile also hunted food for itself but with little success.

Summary

In Costa Rica Scarlet-rumped Caciques are restricted to humid lowlands, where in family groups or straggling flocks they wander through the high forest and clearings with scattered trees. In the evening they gather into larger companies to roost in isolated trees. They bathe in woodland streams.

They search for insects amid the foliage of trees and the epiphytes that burden them, using their feet to hold curled leaves that they probe and large insects that they dismember. They pluck larvae or pupae from wasps' nests and try, sometimes vainly, to extract spittle bugs (cercopids) from the froth that surrounds them. They visit the nectaries of *Marcgravia* and apparently pollinate the ornithophilous flowers. Many fruits and arillate seeds are included in their diet.

These caciques are most vocal, uttering a large variety of loud, bright notes suggestive of urgency and excitement. Females sing much like the males, often while sitting in their nests. Caciques are songful throughout the year.

The male's display before the female, described in detail, is of a type widespread in the Icteridae.

In the Caribbean lowlands Scarlet-rumped Caciques start to nest early in the year and sometimes feed well-grown juveniles by March. These caciques are not colonial and hang their beautifully woven, swinging pouches on exposed branches, from a height of 100 feet to, rarely, as low as 11 feet. Often the nest is close to an occupied vespiary. It is built by the female alone of materials gathered from high in trees, but she is accompanied closely by her mate.

The single accessible nest contained two eggs, white marked with a few light brown and blackish spots and scrawls, which were incubated by the female alone. Six sessions ranged from 18 to 54 minutes in length and averaged 38 minutes. Six recesses varied from 15 to 26 minutes and averaged 20.3 minutes. During six hours this female incubated with a constancy of 65 per cent. Her mate was attentive but he did not enter the nest.

Newly hatched nestlings had pink skin devoid of down and the interior of their mouths was pink like their bodies. At first only their mother fed them but their father was seen to bring food shortly before they were killed by a falling branch.

Whether rapidly or slowly built, nests made in May, apparently in anticipation of a second brood, remained unoccupied for a long while.

Family THRAUPIDAE

WHITE-VENTED EUPHONIA

Euphonia minuta

A small tanager about 3½ inches long, the White-vented Euphonia exhibits the pronounced sexual differences in coloration typical of its genus. The adult male is yellow on the forehead, breast, and much of the abdomen, white on the lower abdomen and under tail coverts, and nearly everywhere else, including the throat, steel blue so dark that it often appears black. There are large, mostly concealed, areas of white on the outer tail feathers, the inner webs of the secondaries, and the under wing coverts. The female is yellowish olive-green above, with a more yellowish forehead. Her cheeks and chin are pale yellowish olive, her throat pale gray, her breast dull yellow, her abdomen white, and her under tail coverts pale brownish gray. In both sexes the bill is grayish with a dusky tip, the eyes are brown, and the legs and toes gray.

The White-vented Euphonia ranges from Guatemala to Peru, northern Brazil, and the Guianas. In northern Central America it occurs only on the wetter Caribbean side and even here has rarely been recorded, but in central and southern Costa Rica and adjacent Panama it inhabits both slopes. Altitudinally it ranges from sea level up to about 5,000 feet in Costa Rica (Carriker, 1910:869) and 3,000 feet in Venezuela (Phelps and Phelps, Jr., 1963:356). It inhabits humid forests, especially their edges and more open parts, and clearings with scattered trees through which it wanders in pairs or small bands, usually remaining well above the ground. Like other euphonias it seems to wander widely, appearing and disappearing in a manner difficult to explain.

White-vented Euphonias seem to subsist on small insects and spiders to a greater extent than many of their congeners. I have often watched them hunting in the treetops on very slender dead twigs, the thickness of a pencil or less. They move outward along the twigs by hopping, by sidling along for an inch or so, and by about-facing, alternately to the right and to the left, by each such reversal of direction bringing the foot that was behind into the forward position. They bend over to examine the lower sides of the twigs, but what they find there is nearly always too small to be detected from the ground. Thus their mode of foraging resembles that which is widespread in the genus *Tangara,* with the difference that these variegated little tanagers bend down to investigate the undersides of thicker branches that are usually mossy or lichen-encrusted. The euphonia varies its procedure by plucking insects from foliage. Like related species it eats many mistletoe berries.

As they hunt over fine twigs the euphonias half-spread their tails and incessantly wag them slowly from side to side. The male's tail then shows a double-rounded posterior margin and displays large areas of white on the inner webs of the outer feathers. The female similarly wags her tail but shows no white. Early in September I watched a male who was molting and had only two full-length rectrices, which gave his little tail a forked aspect. He persistently waved it from side to side, just as though it were whole.

While foraging male White-vented Euphonias often continue to pour forth an artless medley of slight, pleasant notes, a mode of singing not unlike that of certain siskins (*Spinus*). If they have any more formal or forceful song I have failed to hear it. The call of both sexes is a thin, sharp note, constantly reiterated as they go about the business of their nest.

In late July I watched what was evidently a mated pair, of which the male was not yet in full adult plumage; the blue-black areas of his body were flecked with olive. It is probable that in this species, as in certain other euphonias, males may nest in transitional plumage. On 5 April 1965 a male White-vented Euphonia appeared to feed, by regurgitation, another who was evidently his mate. I have also seen nuptial feeding in the Tawny-bellied Euphonia.

Nest and Eggs

The three nests of the White-vented Euphonia that I have seen were all in the Valley of El General at about 2,500 feet above sea level. The earliest was being built at the end of March 1943 and the latest held nestlings toward the end of August 1966. These nests were about 15, 20, and 20 feet above the ground. The first was among the roots and leaf bases of a large orchid plant dangling from a fallen dead branch caught up among vines at the forest's edge, beside a small stream. The second was at the base of a small tank bromeliad growing on a slender mossy branch at the very top of a calabash tree in front of our house. The third and lowest was in a tuft of liverworts (*Frullania*) in an exposed position at the top of a neighboring calabash tree. The last two nests were about 100 feet from the forest's edge.

The first of these nests was built by both sexes. They brought fibrous rootlets plucked from epiphytes high in trees, large billfuls of brownish moss or liverworts, and cobweb. Nearly always the two came together with material in their bills, but occasionally one returned with something for the nest accompanied by an empty-billed partner. When both brought material the male invariably went first to place and arrange his contribution. Then he waited on a nearby dead twig while his mate took her billful to the nest and tucked it into place. After she had finished and returned to perch beside him he would usually go again to the nest, with empty bill, as though to inspect her work. Finally they flew off through the forest together. For 3/4 hour they worked steadily, each bringing material 10 times. Then they remained absent for more than 1/4 hour.

A month later, finding this nest empty, I shook the vine tangle where the nest had been built on an orchid plant attached to a loose dead branch and brought it down for examination. It was of the usual euphonia type, a small, nearly globular structure with a round doorway facing sideways. The exterior was composed of green moss, small living filmy ferns, and orchid roots, all of which blended well with the moss-overgrown mass of orchid roots against which it was placed. In the bottom was a middle layer of seed plumes and long, narrow blades of grass or other monocotyledonous plants. The lining in the bottom was of fine, light-colored plant fibers and some black fungal hyphae or vegetable horsehairs.

I could not learn the contents of the first nest. The second contained three eggs on 29 June. These eggs were white. Two were heavily blotched and spotted all over with brown, which formed an almost solid band around the greatest girth. On the third the brown markings were concentrated on the thick end and more sparingly scattered elsewhere. The third nest held three nestlings in late August.

Incubation

Only the female incubated. I watched the second nest through the morning of 3 July and the afternoon of 6 July, a total of 12½ hours. The euphonia began her active day when she ended her nocturnal session at 05:22 on 3 July, and she settled on the nest to remain through the night at 16:39 on 6 July, as rain began to fall. I timed 17 sessions, ranging from seven to 38 minutes in length and averaging 25.1 minutes. An equal number of recesses ranged from six to 22 minutes and averaged 13.4 minutes. This euphonia incubated with a constancy of 65 per cent. During the first 4½ hours of the morning her sessions were remarkably uniform in length, varying only from 33 to 38 minutes, while her absences gradually lengthened from six-seven to 12-14 minutes. During the remainder of the day both sessions and recesses were more variable.

To leave her closed nest in the exposed treetop, the female euphonia darted out and dropped well below it, to curve upward as she flew away. Often she called after passing beyond view. Although I rarely saw or heard her mate while she incubated, evidently she knew where to find him for on 11 of her 18 returns he came with her. The two alighted on a slender dead twig about four feet from the nest and somewhat above it. After a pause here they started to fly together to the nest, as though racing to reach it first. The female always passed through the doorway. The male stopped in the air above the nest and hovered there a moment, or else he alighted on the roof or, more rarely, close beside it. In any case, he promptly flew away and I saw no more of him until he accompanied his partner on her return from her next outing. Whenever the female returned alone she called much with sharp, high-pitched notes and low *chip*'s before she entered the nest. When she returned with her mate she was mostly silent, although he usually uttered high and low notes.

On 16 July, after at least 17 days of incubation, the three eggs hatched. The incubation period of euphonias is substantially longer than that of most other tanagers. The empty shells were promptly removed.

Nestlings

Development.—The newly hatched euphonias were covered with shiny, blackish skin that was nearly naked, with only a few short tufts of gray down on the hindhead, shoulders, and center of the back. Their eyes were tightly closed. The interior of their mouths was red. Their development was much slower than that of most tanagers. At the age of nine days their eyes were open and the pins of the remiges had become long, but the rudiments of the other feathers barely protruded through the skin. As in many other nestlings that are fed by regurgitation there was a swelling on the front and right side of the neck, caused by a dilatation of the oesophagus that became large and prominent when distended by a meal. Even at 11 days, when many larger tanagers are feathered and ready to leave the nest, these young euphonias were nearly naked. When they were 13 days old the feathers on their backs and wings began to expand and at 15 days their dorsal surface, with the exception of the head, was fairly well covered with greenish plumage. They remained in the nest for five days more.

Brooding.—The nestlings were brooded very little. On 21 July, when they were five days old and still quite naked, I watched for five hours of the morning. Their mother, who had passed the night with them, flew from the nest at 05:15. During the next three hours she brooded six times, for intervals ranging from five to 14 minutes and totalling 61 minutes. From 08:15 to 10:15 on this morning, however, she did not brood at all; nor did I see her do so on subsequent days, no matter how hard it rained. The male, who had not incubated, never brooded.

The female parent continued to brood by night until the nestlings were eight days old after which, despite their continuing nudity, they passed the often chilly nights without a cover. Although the roof of their nest afforded some protection from the frequent rain of this season, this seemed rather harsh treatment. But I recalled that, at this same altitude, young hummingbirds, also fed by regurgitation, likewise pass the night alone before they are feathered, in open nests that fail to shield them from the rain. When only six days old a brood of Black-faced Grosbeaks was left alone at night in an open nest roofed above by epiphytes (see p. 216).

Feeding.—Throughout the nestling period, the two parents nearly always arrived together to feed their young. There was never any food visible in their bills. The male invariably fed the nestlings first; even when his mate arrived a little before him she perched nearby and called until he came. She continued to rest near the nest, repeating sharp notes, during the rather protracted business of feeding by regurgitation. After he had finished and alighted close to her, she went to deliver her food while he waited. When she did not stay to brood, as after the nestlings' first week she never did, the two parents always flew off together at the conclusion of the feeding.

During the nestlings' first days the female sometimes seemed to feed them after entering the nest, but I could not see enough to be certain of this. The male always fed while standing in front of the nest, as did the female after she ceased to brood. The time spent delivering a meal decreased as the nestlings grew older. When they were five days old the male spent from 50 to 94 seconds with his head in the doorway, but I could not learn how much of this interval was occupied by regurgitation to the invisible nestlings. After they grew bigger I could follow the details of feeding. When a parent alighted in front of the nest three mouths opened like red flowers before it. The parent proceeded to regurgitate food and place it in these corollas a little at a time, first in one and then in another. The parent's short bill was not inserted into the nestling's throat and there was no long-continued regurgitation with evident muscular straining, as in hummingbirds. When these nestlings were 15 days old I timed a number of feedings with a stop-watch. Seven feedings by the male, whose position while delivering food was more favorable for observation than that of his mate, continued from 13 to 35 seconds, during which from four to 12 separate portions of food were brought up and passed to the nestlings. Once the female delivered 11 portions in 30 seconds, and on another visit 14 portions in 35 seconds. A single nestling might get several portions in succession, but all three of them seemed to receive food on nearly every one of the parents' visits. As well as I could see, each portion consisted of a number of small particles which were deposited in the nestling's mouth with rapid movements.

During the first five hours of the fifth morning after the nestlings hatched, the male brought food nine times. The female fed three times from outside the nest and possibly several times more after she entered to brood, when her movements were difficult to follow. In the corresponding period of the twelfth morning after the nestlings hatched, the parents came together with food nine times, making 18 feedings. From 13:00 to 18:00 on 31 July, when the nestlings were 15 days old, there were also nine feeding visits by both parents in five hours. Thus through most of the nestling period the parents came together with food at intervals that usually fell between 30 and 40 minutes.

Sanitation.—At the conclusion of a meal one or more nestlings turned right around and raised the posterior end to defecate. The dropping was not neatly enclosed in a white fecal sac, as in most passerine nestlings, but appeared to be rather gelatinous. The parent ate it little by little, seeming almost to lick it up. I never saw them carry anything away in their bills. Sometimes the fecal matter pulled out into long, clear strings, like the droppings of adult euphonias. This colorless, viscid substance was probably from the seeds of mistletoes (Loranthaceae), a favorite food of many species of euphonia; it serves to stick these small seeds to the branches of trees where they germinate. Although I never saw this pair of euphonias eat mistletoe seeds, doubtless they gathered quantities of them at a distance.

Defense of the nest.—Sometimes when I climbed the long ladder to inspect the nest the female euphonia alighted on a naked twig at the very top of the nest tree, a few feet from me. Here, wildly flitting her wings, she poured forth a cascade of notes in strongly contrasting keys, first a rapid sequence of rather low notes, next an equally rapid series of high, thin notes. Then she dashed away, perhaps to return promptly and repeat the performance. The male usually held aloof while I was near the nest.

One morning, when the parents were absent collecting food for the nestlings, another pair of White-vented Euphonias alighted near the nest and the male pulled something small from the moss and lichens that covered the calabash tree. When the resident pair returned and found these intruders they flew at them, the male parent at the male stranger, the female at the female. These pursuits were mild; the birds did not clash together; and after a little chasing the four would perch for a while not far apart in the nest tree. Only after the trespassers flew away did the parents proceed to feed their nestlings. Later in the morning the same or another pair of invading euphonias came to the nest tree and were chased when the parents returned to feed the nestlings; but this time the strangers did not stay long.

When a female Tawny-bellied Euphonia looked intently into the nest, the parents did nothing but complain a short distance away.

Departure.—The three young euphonias flew from their nest before 07:30 in the morning of 5 August, when they were 20 days old. I found one in a guava tree near the nest, perching on a high, thin twig, where its father fed it. It preened much, then dozed. In plumage it resembled its mother but it had scarcely any tail and there were prominent fleshy knobs at the corners of its mouth. These fledglings did not return to sleep in their nest.

Summary

White-vented Euphonias wander widely through the upper levels of the forest and clearings with scattered trees. Much of their food consists of tiny insects and spiders that they glean in characteristic fashion from thin dead twigs in the treetops. While foraging they constantly wag their tails sideways. Like other euphonias they eat mistletoe berries.

The male's song is an artless, rambling medley of slight, pleasant notes.

In the Valley of El General these euphonias breed from late March to August. Three nests were situated at heights of 15 to 20 feet among the roots of an epiphytic orchid or on branches heavily swathed with mosses and liverworts. The globular, roofed nest, with a round doorway in the side, is built by both sexes. The two usually come together with material and the male always places his contribution before the female.

The single accessible nest contained three eggs, which were white, heavily blotched and spotted with brown.

Only the female incubated, taking sessions that ranged from seven to 38 minutes and averaged 25.1 minutes, separated by recesses which ranged from six to 22 minutes and averaged 13.4 minutes. During 12½ hours she sat with a constancy of 65 per cent. As she returned to her nest her mate escorted her closely as though racing her to the doorway, a custom widespread among euphonias.

The incubation period was at least 17 days.

The newly hatched euphonias had only a few short tufts of gray down on their shiny blackish skin. The inside of their mouths was red. They developed much more slowly than most tanagers and were not feathered until they were about 15 days old.

Only the female brooded and she did so very little, leaving the young alone at night when they were only eight days old and still practically naked. This early cessation of nocturnal brooding is compared with the similar habit of hummingbirds and certain other birds.

The nestlings were fed by both parents, always by regurgitation. Nearly always both arrived together and the male invariably preceded his mate in delivering food.

The nestlings' gelatinous droppings were not enclosed in white fecal sacs and seemed almost to be licked up by the parents, who never carried anything away in their bills.

The nestlings left the nest at the age of 20 days, when they could fly well. They did not return to sleep in the nest.

TAWNY-BELLIED EUPHONIA

Euphonia imitans

The Tawny-bellied Euphonia is a small tanager nearly four inches long, and as in most members of its genus, the sexes differ greatly in coloration. In the male the forehead and crown are lemon yellow, with partly exposed central triangular spots of black on their feathers. The remaining upper parts are glossy blue-black or dark steel blue, as are the sides of the head and neck, the chin, throat, and upper chest. The remiges are black with bluish green margins and much white on the usually concealed basal portions of most of them. The black tail feathers have bluish green margins but are devoid of white. The more posterior under parts are rich lemon yellow. In the female the forehead is rufous-chestnut and the remaining upper plumage is olive-green glossed with metallic bluish green. The wings and tail are dusky with yellowish olive-green edgings. The sides of the head and neck and most of the under parts are olive-green with a yellowish tinge, especially on the throat, and the abdomen and under tail coverts are tawny-ochraceous. In both sexes the short, thick bill is largely black; the eyes, legs, and toes are dark.

The male of this species closely resembles the male Yellow-crowned Euphonia which occurs in the same region. The chief differences are the greater backward extension of the yellow on the top of the latter's head and its lack of black spots. But these characters are so difficult to detect in the field that one must usually depend on the very different notes of these similar species in order to identify them with certainty. The female Tawny-bellied Euphonia is, however, readily distinguished from the female Yellow-crowned Euphonia by her richer and more varied colors, especially the chestnut of her forehead and the tawny-ochraceous of her abdomen, both of which colors are absent from the female Yellow-crowned Euphonia. Since in the field we depend largely on the female to identify *Euphonia imitans,* it seems appropriate that in this case we depart from the usual practice of naming a species from a character of the male rather than of the female.

The Tawny-bellied Euphonia has a restricted range on the Pacific slope of Costa Rica south of the Gulf of Nicoya and in neighboring parts of Panama. An old record of its occurrence on the Volcán Miravalles in northwestern Costa Rica is dismissed as erroneous by Slud (1964:349); and a specimen collected long ago near San José may have been an escaped cage bird. Altitudinally this euphonia ranges from sea level up to about 4,500 feet. Nowhere have I found it more abundant than in the Cañas Gordas district between 3,500 and 4,000 feet, and it is also fairly common in the remaining forests of the basin of El General up to at least 4,000 feet. The Tawny-bellied Euphonia is primarily a bird of the heavy forest, whence it makes excursions into neighboring clearings with scattered trees; whereas the Yellow-crowned Euphonia, common in the same region, is a bird of the clearings that occasionally visits the crowns of the forest trees. Both species have nested in our dooryard, which adjoins the forest. The Tawny-bellied Euphonia travels alone or in pairs and seems never to flock, as some other kinds of euphonias do.

Food

Tawny-bellied Euphonias search for insects and their larvae on mossy branches and among dead leaves caught in tangles of vines well above the ground. They eat the green fruiting spikes of shrubs and small trees of the genus *Piper*, so abundant in the forested regions of tropical America, and the berries of the equally common melastomes. When they take the berries of the epiphytic solanaceous shrub *Lycianthes synanthera* they press and bite out the pulp, then drop the tough, greenish yellow skin, just as other tanagers do. Occasionally a Tawny-bellied Euphonia comes to our feeder for bananas, as does the Yellow-crowned Euphonia; but neither of these species is a regular attendant.

One morning in late December I stood in a riverside thicket, among tall slender-stemmed shrubs of the acacia-like *Calliandra similis*, which were flowering profusely. The numerous, long, slender, red stamens sprang in a cluster from the little, five-toothed, green calyx; there was no corolla. As the rays of the rising sun pierced brightly down into the thicket, these tiny green chalices were full of sweetish liquid, which if not pure nectar was nectar mixed with water from the rain that had fallen during the past night; and a pair of Tawny-bellied Euphonias were visiting them. Even their short, thick bills could reach the sweet fluid. They seemed also to squeeze the calyces in their bills, to press out the nectar that they could not reach with their tongues. With them was a Golden-masked Tanager also sampling these sweets.

One morning I watched a Streaked Saltator digging into a full-grown but still green guava (*Psidium guajava*) in the top of a tree behind our house. It pecked out small bits and chewed them, dropping fragments from its bill. Meanwhile, a male Tawny-bellied Euphonia hopped around nearby, waiting his turn at the fruit. He gathered from the leaves particles that the saltator had dropped and ate them. Becoming impatient, he clung to the other side of the saltator's fruit and started to eat, but was driven off by a peck. After lingering close by a little longer, he again clung to the guava, and this time the saltator permitted him to share it. The saltator soon flew off but the euphonia did not remain in possession of the fruit for long because the former bird's mate, as I judge, who had been waiting nearby, now claimed it. After the second saltator had satisfied itself the euphonia returned to the fruit and ate freely, pecking out bits from the inside and mandibulating them, just as the saltators did. But before the tiny bird had finished, a Blue Tanager took the fruit from him. The tanager ate little and when it left the euphonia resumed his meal. Soon, however, he was displaced by a Buff-throated Saltator and flew away. The euphonia had spent much more time waiting than eating.

Once while a female Tawny-bellied Euphonia hunted over a slender vine in the forest, a male alighted beside her and gave her food. At times the female may return this courtesy. One day I watched a female pluck a long, gray caterpillar from some dead leaves caught up in a tangle of vines. Before she could swallow her prize, a male of her kind, in full nuptial plumage, flew up and took it from her. Although it was not clear that she voluntarily relinquished the caterpillar, at least she did not try to withhold it from him. I have seen only these two instances of nuptial feeding in the Tawny-bellied Euphonia. The male fed the female in early March of 1942; the female fed the male in early June of the following year.

Sleeping

Tanagers with open nests sleep amid foliage, the male and female of a pair often roosting from a foot to a yard apart rather than in contact with each other, in the few species that I have seen by night. As is to be expected of a bird that raises its family in a covered nest, the Tawny-bellied Euphonia seeks a more sheltered place for sleeping. In front of our house at Los Cusingos grow several calabash trees, with stout, stiff, leafy branches densely covered with ferns, orchids, lichens, mosses, and thick, dirty-brown cushions of the liverwort *Frullania*. For the past five years, euphonias have slept intermittently in sheltered nooks in these small trees. I first discovered one of them lodging here on 22 September 1966 when I found a male in full adult plumage in a small pocket among green moss that hung in a dense mass below a horizontal branch, about 15 feet above the ground. This niche had a sideward-facing opening and was of a form that, with adequate lining, might have made a typical breeding nest. I do not know whether the euphonia shaped the pocket in the moss or found it waiting for him. He continued to sleep in it until I left in mid-October. When I returned a month later he no longer lodged here, and I have no other record of him until November 1967, when for a few nights he slept in a similar pocket amid the moss and liverworts.

In late October 1968 the male euphonia returned to sleep in one of these calabash trees. After searching over the trees on several mornings he selected a different kind of roost, a slender, cord-like rhizome of a creeping fern, which passed loosely beneath a branch swathed in small epiphytes. The space between the rhizome and the branch above was so narrow that the euphonia perched with his back pressed against the underside of the latter, at a point where a luxuriant growth of *Frullania* stood out broadly on either side, forming a wide, spongy, brown roof above him. With his head turned back and his ventral plumage all fluffed ont, he looked, in my flashlight's beam, like a little ball of bright yellow feathers. He clutched the fern rhizome with one foot; the other was drawn up and hidden in the yellow mass. I thought that the bird might have difficulty inserting himself into the narrow space between rhizome and branch, but he slipped in so swiftly that I could not see how he did it.

After sleeping here for a few weeks the male euphonia found more distant quarters. I did not notice him roosting again until early in 1971. Now he slept upon the thin root of an epiphyte that formed a loop beneath a horizontal branch, with his yellow under parts wholly exposed but shielded above by densely matted liverworts. With occasional inexplicable absences he used this shelter for the next four months. Sometimes he came silently to his dormitory but on other evenings he rested in a neighboring tree and for many minutes poured forth an almost continuous stream of chaffy, rattling, and less often liquid, notes before he retired for the night. The time when this occurred varied considerably from evening to evening. In the dim light of dawn he darted silently out and away.

In February 1968 I first noticed a female sleeping in a pocket amid the liverworts, similar to that in which I first found the male. After trying several niches she found one so satisfactory that she continued to lodge in it for nearly two years, with occasional absences. This dormitory was 10 feet above the lawn, far out on a horizontal branch in a large tuft of brown liverworts hanging beneath the limb. Often after nightfall I have thrown up the beam of a flashlight and beheld her tawny breast filling the round opening in the side of the snug niche, into which she just fitted. On many nights the male, evidently her mate, roosted on the root beneath the same branch about two feet away.

The female euphonia's absences from her niche were sometimes inexplicable but at other times they were caused by a Bananaquit. These diminutive, yellow-breasted honeycreepers are among the most indefatigable of nest-builders, seeming to spend about half their lives making cozy, covered structures, both for their eggs and young and for sleeping, always one adult to a nest. Despite their great industry, evening sometimes finds them without a roof of their own and then they try to enter the covered nest of some other small bird, often a flycatcher. One evening in the calabash tree where the euphonias slept, I watched a Bananaquit try persistently to intrude into the little globular nest where a tiny Paltry Tyranniscus was incubating, but the distressed parents managed to drive the interloper away. One day last January I noticed that a Bananaquit had arranged fresh straws around the doorway of the pocket where the female euphonia had already lodged for many months, reducing the aperture. In the evening when the euphonia arrived to enter her dormitory, she paused clinging in front, repeating her peculiar *churr*. Trying to enter she was greeted by pecks from the small bird inside. But she persisted in pushing in until she and the Bananaquit tumbled out, locked together, and fell to the ground. Separating, they flew up to the pocket together and again they clutched and dropped to the lawn. Finally the euphonia returned alone and remained for the night. But afterward the Bananaquit added so much material to the front of the pocket, making the entrance face downward rather than sideward, that the euphonia relinquished it to the tireless builder.

Once, in early August, I watched a female Tawny-bellied Euphonia cling to a tuft of liverworts and mosses hanging from a high branch and tear out pieces that she dropped, making a pocket in the moss. As far as I could learn, this pocket was never used for either sleeping or nesting but her activity suggested that these euphonias may themselves prepare the niches amid epiphytes in which they sleep or build their nests.

Probably other species of euphonias likewise sleep in sheltered pockets, but the Tawny-bellied Euphonia is still the only tanager that I have found roosting elsewhere than amid foliage.

Voice

The Tawny-bellied Euphonia sings most freely in the sunny days at the beginning of the dry season early in the year, when the air is still fresh and clear and flowers abound. Sometimes the male perches inconspicuously in a tree and repeats interminably, for a quarter or even half an hour without interruption, various combinations of short notes, which may be either clear or dry and chaffy. At times the clear whistles are grouped in doublets and trios. Or the verse may consist of four rather dry notes grouped in pairs. One male sang *chip a cher weet,* with the first and last syllables accented. A series of drier notes sounded like *chip tuck tuck,* with the emphasis on the *chip.* Far from a brilliant songster, the euphonia performs with more enthusiasm than talent, yet occasionally he achieves a delightful, long-continued medley that includes a variety of low, soft notes and warbles, in addition to his usual chaffy and burry sounds.

While the male pours forth his artless song for minutes together, the female may answer with a dry, chaffy *churr.* Sometimes she sings somewhat in the manner of the male, when no companion is in sight. When her nest is disturbed she may continue to call as long as the intruder stays in view. The rapid, throaty rattle she then utters is interspersed with higher, chaffy notes, or with a sharp *chip chip chip.*

The notes of the Tawny-bellied Euphonia have little or none of the metallic twang that is so pronounced in the Yellow-crowned Euphonia's song; and its *chip tuck tuck* is quite different from the call of the latter, which sounds like *be be be.* After one has become familiar with their voices he is not likely to confuse these two similar species where they occur together in southern Costa Rica.

Nest Building

The 10 nests of the Tawny-bellied Euphonia that I have seen on our farm in El General and near Cañas Gordas, fall into two groups. Six of them were on the sides of erect, mossy trunks or ascending mossy branches, at heights of 6½ to 25 feet. Of these six, four were in the midst of the forest and two, in which I never found eggs, were built on calabash trees in our dooryard. The first of the nests on mossy trunks or branches was set among the roots of an epiphytic aroid; the second in the midst of a cluster of small orchids and moss; the third above a tuft of small hart's-tongue ferns (*Elaphoglossum* sp.); the fourth was simply embedded in the thick coat of moss on a slender trunk; the fifth was at the base of an epiphytic orchid; and the sixth, the only one on an ascending branch rather than an erect trunk, was sheltered beneath a large tank bromeliad.

The other nests were in hanging sites. Six feet above the rocky channel of the Río Peña Blanca, far out from shore, beneath a far-reaching horizontal limb of a great sotacaballo (*Pithecolobium longifolium*) tree, hung a segment of a lateral branch that had long before died and broken away but did not fall into the river because it was bound to the rest of the tree by the cord-like roots of epiphytes. This short piece of branch hung upright and was everywhere covered with green moss, in which small ferns, orchids, and other epiphytic plants had rooted. The nest, built near the bottom of this aerial garden, appeared to be a thicker growth of the moss that enveloped it. Composed largely of living vegetation, this nest persisted through the long rainy season with scarcely any deterioration and was used again in the following year. It was about 100 yards from the forest. In a wooded ravine near Cañas Gordas I found a nest similarly situated at the lower end of a dangling, epiphyte-covered segment of dead branch, which in this instance hung about 20 feet above the bank of a narrow brook instead of low above the middle of a wide mountain torrent. Yet another swinging nest was 11½ feet above the ground in an elongated tangle of epiphytic roots that hung below a thick, horizontal limb of an old, epiphyte-laden tree at the end of our garden, hard by the forest.

As in other euphonias, both sexes of the Tawny-bellied Euphonia build the nest taking rather equal shares in the task. This was true of three pairs that I watched at work. They gathered their materials from air plants growing high on the trees and plucked cobweb from the foliage. I did not see one partner pass its billful to the other for placement in the nest, but each worked its own contributions into the structure. If it found its mate in the nest when it arrived it waited nearby, holding its load, until the other left. In 40 minutes a building male entered the nest 19 times and his partner did so 17 times. I did not notice material in the euphonia's bill each time it went into the nest; some of these visits were apparently specially for shaping it. At one nest building continued for five or six days, but not always at the same rapid pace.

The completed nest is a globular, roofed structure with a round doorway in the side. This orifice is sometimes well shielded by a visor-like projection from the roof. One nest was composed of fine, richly branched rootlets and green moss, and lined with fine fibrous material. Another nest was made of much green moss and a few small plants of the epiphytic fern *Rhipidopteris peltata,* with a lining of narrow strips of bark, many dark fibrous rootlets, some shiny black strands of the fungous growth known as vegetable horsehair, and some light-colored fibers. One nest was about 3½ inches in outside diameter and the round doorway in its side was 1½ inches wide.

Eggs

My earliest date for eggs is 29 March 1948 when a set of three was completed. They are laid on consecutive days. Each of four nests contained three eggs and two held two eggs. The eggs are long ovate in form. They are white or pinkish white, speckled and blotched with bright reddish brown, cinnamon, or chocolate, usually thickly on the broader end and sparingly over the remaining surface, where the flecks of color may be finer and paler. The measurements of 13 eggs average 19.7 by 12.8 mm. Those showing the four extremes measured *21.4* by *13.1, 18.1* by 13.0, and 20.2 by *11.9* mm.

In southern Costa Rica (El General and Cañas Gordas), 2,500 to 3,700 feet above sea level, eggs were laid in seven nests as follows: March, 1; April, 5; May, 1. A nest was built late in June in an inaccessible site on a calabash tree in front of our house, but I never found the female incubating in it.

Incubation

As in other tanagers only the female incubates. On 23 April 1942 I spent four hours of the morning watching a nest situated on the side of a mossy trunk in the forest. The female always sat in her snug little chamber facing outward, with her chestnut forehead visible from the ground. She incubated with great patience; the two sessions that I timed lasted 78 and 55 minutes; the three recesses 30, 38, and 27 minutes. To leave her nest she habitually dropped from her doorway like a dead weight and fell almost to the ground before she veered upward and rose into the trees. This mode of departure is frequent among small birds with high nests. Its value seems to be that it tends to deceive an enemy that has the nest in its field of vision but has not yet detected it. The departing bird's flight often appears to originate at the point where she breaks her fall and turns upward, for here her movement is slowest, and it is here that the nest-robber might search for her nest—fruitlessly. Whether the euphonia left her eggs spontaneously or fled when I set up a ladder to climb up and inspect them, she always departed with the initial drop.

Each time that she returned the female was accompanied by her mate, who alighted with her among the branches of a neighboring low tree while she surveyed her surroundings before approaching her nest. When finally she flew to her doorway he pursued her so closely that he seemed to be racing to reach it before her. But just in front of the nest he veered aside and left her to enter alone. This mode of escorting the female to the doorway is widespread among small birds whose closed nests have round openings in the side. I have also seen it in the Yellow-throated or Bonaparte Euphonia, White-vented Euphonia, Golden-browed Chlorophonia, Black-fronted Tody-Flycatcher, and other species.

Fifteen years later, in the same forest, I watched another nest in which incubation was in progress. My vigils extended from 05:25 to 11:13 on 28 April 1957 and from 12:15 to 18:00 on the following day. Both the morning and the afternoon were sunny. On this afternoon the female returned from her last excursion and settled down for the night at 16:15, nearly two hours before it became dark in the forest, so that my record covered only about 10 hours of her active period. This euphonia also sat with great patience. The five sessions which I timed varied from 77 to 108 minutes and averaged 86.8 minutes. Her five recesses ranged from 13 to 43 minutes and averaged 29.8 minutes. She covered her eggs for 74.4 per cent of the 10 hours.

After darting into her nest the euphonia turned around in a trice and settled down facing outward; she always preserved this orientation throughout her long session. When she left she also seemed to fall from her doorway, to drop straight down beside her mossy trunk for nearly 20 feet before she curved sharply to fly up into the trees from a point near the ground. Her mate stayed out of sight most of the time although I often heard, coming from the treetops above me, notes that I attributed to him. On the two occasions when he accompanied the female to the nest he seemed to race with her as she flew the final few feet, just as I had seen at the first nest.

The Tawny-bellied Euphonia is not easily frightened from her eggs. At the highest of the hanging nests the female continued to sit while I made her dangling, epiphyte-covered segment of branch swing by shaking the slender tree from which it hung. Whenever I visited the nest hanging low above the channel of the river the euphonia would watch me approach slowly over the rocks until I looked in at her, with my eyes only a yard or so from hers. Finally, at some slight movement of mine, she would dart out and rise to a bough above the water where she repeated interminably a throaty rattle followed by higher, chaffy notes. One day she continued to complain in this fashion the whole time that I bathed in a neighboring pool. Her mate did not appear. A female incubating in a low nest embedded in the moss that covered a trunk beside a forest trail was even bolder, returning my gaze when my head was hardly a foot away. If I advanced a hand she would drop into the undergrowth beside the trail, calling *chu chu chu chu* very rapidly. Then, screened by the foliage near the ground, she repeated over and over a loud phrase that sounded like *wer de d'der*.

At this nest one egg hatched 18 days after the set of two was completed. This seems a long incubation period for a tanager; but at a nest of the White-vented Euphonia the period was 17 days or more, and at a nest of the Yellow-throated Euphonia it was 16 days.

In five other accessible nests the eggs did not survive until they hatched. I do not know what destroyed them except in one instance, when the culprit was a squirrel. One day as I approached this nest I heard the female chattering continuously. A squirrel was nearby and two fresh eggs lay broken on the ground below. I chased the rodent away and for the next quarter of an hour the euphonia continued to utter her rapid throaty *churr,* which at intervals she punctuated with a sharp *chip chip chip.* As she called she flipped out her short wings simultaneously and turned from side to side. Finally she started to eat the greenish fruits of *Topobea Durandiana,* an epiphytic shrub of the melastome family that grew on the tree that supported the nest. Then her mate arrived and she flew away with him. One of her three eggs remained intact and later she returned to incubate it, but after a few days it vanished.

The one nestling that hatched in the nests over which I kept watch also disappeared, the day after it escaped from the shell. Hence I lack observations on the care of the young in this species; but without much doubt both parents feed them by regurgitation and they remain in the nest a long while, as in other euphonias. A brood of three White-vented Euphonias that I recently studied did not leave their roofed nest until they were 20 days old and Yellow-crowned Euphonias remain in the nest until they are from 22 to 24 days old (Skutch, 1954).

Summary

On the Pacific slope of southern Costa Rica the Tawny-bellied Euphonia ranges from sea level up to about 4,500 feet. It inhabits the forest, whence it makes excursions into neighboring clearings with scattered trees, alone or in pairs, never in flocks.

These euphonias search for insects and spiders on mossy branches and among dead leaves caught in vine tangles. They eat a variety of fruits, and sip nectar from shallow calyces. The male occasionally feeds his mate and sometimes he receives food from her.

Tawny-bellied Euphonias sleep in sideward-facing pockets amid moss and liverworts on the limbs of trees, or resting on an epiphytic root or rhizome close beneath a dense mass of bryophytes that shields them from rain. Sometimes they must compete with Bananaquits for these snug dormitories.

The song, usually a mixture of short whistles and dry, chaffy notes, may continue uninterruptedly for many minutes. Rarely it becomes a delightful medley of low, soft notes and warbles, mixed with less liquid sounds.

The nest, a globular structure with a side entrance, may be built in the forest or a neighboring clearing. Usually it is embedded in moss that densely covers a trunk or ascending limb which also supports larger epiphytes, but it may be attached to a hanging tangle of epiphytic roots or to a length of dead branch that dangles from such a root. The sexes take almost equal shares in building, bringing much green moss, fine rootlets, small epiphytic ferns, and fibrous materials for the lining.

Laying begins in late March and continues at least until May. The set consists of two, or more often three, white or pinkish white eggs, speckled with brown or chocolate.

Only the female incubates, sitting steadfastly for intervals that usually exceed one hour. To leave her nest she falls straight downward almost to the ground before turning upward into the trees. As she returns to her eggs her mate often escorts her so closely that he appears to race her to the doorway.

In one instance the incubation period was 18 days. The nestling period is unknown.

BLUE TANAGER

Thraupis episcopus

In an earlier work (Skutch, 1954), I told what I knew about the habits of the widespread, versatile Blue or Blue-gray Tanager. The purpose of the present chapter is to add certain more recent observations, especially those at a nest attended by a polygamous trio—one of the very few known instances of polygamy in the tanager family.

Food

In my earlier account I mentioned the Blue Tanager's fondness for a variety of fruits and described its methods of finding insects on leaves and branches. More recently, Snow and Snow (1971) have shown that this tanager's procedures on the island of Trinidad are quite similar to those that I have watched on the continent. Of special interest is their description of how it deals with the long, slender, green, fruiting spikes of species of *Piper,* which resemble those of Cecropia trees but are thinner and often upright rather than pendent. The bird breaks off a spike, carries it away, lays it across a suitable branch, and picks pieces from it.

For the last decade a pair of Blue Tanagers has been coming frequently to eat leaves of *Cassia indecora,* a straggling native shrub that, uninvited, scrambled to the top of a poró (*Erythrina berteroana*) tree beside our house and has been preserved because of the profusion of yellow flowers that adorn it early in the dry season. The tanagers tear off small pieces of the leaflets, mash them in their bills, and swallow them, as we have repeatedly seen. Many of the compound leaves are partly eaten away, testimony of the tanagers' frequent visits. We say that the birds come for their salad. My wife has watched a female Scarlet-rumped Black Tanager eat these same leaves, which to me have little taste. A number of tanagers and finches devour the young leaves and shoots of vines. In Venezuela I watched a pair of Grayish Saltators consume a large quantity of the tender terminal shoots of a cucurbitaceous vine, passing pieces to a juvenile, as big as themselves and well able to feed itself. Streaked Saltators, Buff-throated Saltators, and apparently also Blue Tanagers eat young shoots, as well as the green flowers and swelling ovaries, of the chayote (*Sechium edule*) and tacaco (*Polakowskia tacaco*), decreasing the productivity of these cucurbitaceous vines cultivated for their edible fruits.

More surprising is the fact that, at a hotel on the island of Tobago, Blue Tanagers along with White-lined Tanagers and Red-crowned Woodpeckers "came steadily to the sugar dishes from early morning to dusk." Such addiction to sugar has long been widespread among Bananaquits in the West Indies but seems not to have been previously recorded of the Blue Tanager. Apparently the tanagers developed their taste for sugar after the hurricane of September 1963, which stripped much of the foliage from trees and shrubs and made food so scarce that birds thronged about houses in search of it (Hundley and Mason, 1965). Insular Blue Tanagers are certainly much more confiding in the human presence than their continental cousins; everywhere on the mainland I have found the Blue Tanager a very shy, distrustful bird.

Sleeping

In my earlier account I told that Blue Tanagers roost amid dense foliage, often in orange trees, where they may rest on long thorns, the members of a pair sleeping a foot or more apart, never in contact. This has been confirmed by later observations. From March until July 1953 a pair roosted about eight feet above the ground among the outermost foliage of an orange tree behind our house. Sometimes they were separated from each other by about their own length; they were never in contact. Some years later another pair was found roosting on low branches of an orange tree about two feet apart. I have never seen tanagers of any kind sleeping pressed together but often the members of a pair are not far apart.

Voice

Like many wide-ranging birds the Blue Tanager has developed local song dialects. Near Pirapira, state of Carabobo, northern Venezuela, I noticed that the abundant Blue Tanagers' song started off much like that of Costa Rican Blue Tanagers but it ran off into high, wiry notes like those of the Bay-headed Tanager—notes that I have not heard from the Blue Tanagers in Central America. The resemblance of these phrases to those of the Bay-headed Tanagers about my home in Costa Rica was so great that I sometimes looked for this widely distributed bird, but I failed to find it in the locality where the Blue Tanagers used such confusingly similar notes.

Anting

When I wrote my earlier account I had no record of anting by Blue Tanagers, but since then I have twice watched them engage in this puzzling activity. As in every other instance of anting that I have witnessed in tropical America, involving birds of a number of families, it was done in a tree rather than on the ground, as is usually, but by no means invariably, the way in the North Temperate Zone. On the morning of 29 November 1958 a Blue Tanager hopped from twig to twig of a lemon tree until it found something that I could not distinguish. Then it brought a wing forward in front of its breast and rubbed its bill along the inner surface. As it did so its tail was bent down and forward as far as its perch permitted. The tanager repeated this act many times during several minutes.

On 4 December 1969 a Blue Tanager picked small objects from the trunk and thicker limbs of a tall Cecropia tree, then rubbed these unseen objects beneath a slightly raised wing, repeating this sequence numerous times. It was evidently anting with the small, harmless Azteca ants that inhabit the hollow trunks and branches of swiftly growing Cecropia trees.

Nesting

Although Blue Tanagers usually build their compact open cups in a tree or shrub, from near the ground to 100 feet up, in Costa Rica they sometimes place their nest beneath the thatched roof of a shed, as I earlier recorded. Not only do they take advantage of structures made by man, they may on occasion seek the shelter of the large nest of some other bird that builds more elaborately. In Venezuela in 1966 I found three pairs of Blue Tanagers carrying material into as many nests of Rufous-fronted Thornbirds, massive edifices constructed of interlaced sticks and containing several chambers, each of which has its own doorway in the side. While persecuting the wren-sized, brown builders of one of these swinging nests (who took refuge inside a chamber) the Blue Tanagers paid no attention to my presence below them, although afterward, when in a less aggressive mood, they exhibited their usual wariness and would not continue to build even when I watched from a good distance. In Venezuela, as I have seen also in Guatemala, Costa Rica, and Peru, both sexes bring material to the nest.

I could not tell whether the tanagers laid in any of the three thornbirds' homes but later, when one of these structures fell, I found a well-made tanagers' nest in the lowest chamber, which had been falling apart and exposing the interior, so that the tanagers had easy access to their nest while they enjoyed the shelter provided by the whole massive structure above them. In an upper chamber of this same structure three thornbirds continued to sleep, undisturbed by the intruders below them. (Skutch, 1969:135-136).

Even here where thornbirds' hanging nests were so numerous and conspicuous, most Blue Tanagers built amid the foliage of trees and shrubs. In the same locality where I noticed three pairs intruding into thornbirds' homes, I found five pairs nesting in the more typical open situations. Ranging in height from seven to 25 feet, these nests were neat open cups. One was composed of petioles, fibers, bits of papery leaf-sheath, spiders' egg cases, and the like. On the outside of another was much fluffy white cotton from neighboring wild cotton plants. Between 7 April and 13 July 1966 each of these nests contained two eggs, ranging in color from pale blue-gray to nearly white, all mottled and flecked with shades of brown, which on the thicker end of one nearly obscured the ground color. Another egg bore a number of coarse blackish spots. The eggs of one set measured 23.2 by 16.9 and 22.3 by 17.0 mm; of another, 25.2 by 16.8 and 24.7 by 17.0 mm.

Although I earlier mentioned that in Costa Rica Blue Tanagers range upward from the coasts to 7,500 feet, I had seen no nest at such high altitudes. On 16 May 1963 I found a pair building at about 7,300 feet on the Barba massif in the Cordillera Central of Costa Rica and on 20 June I examined a nest with two eggs, at 6,800 feet in the same locality.

When I wrote the earlier account I had no record of nesting in February or August in the Valley of El General, Costa Rica, where chiefly I studied this bird. Subsequently, I found a single nest in which eggs were laid in each of these months. Combining earlier with later records, in 62 nests in this valley eggs were laid as follows: February, 1; March, 16; April, 28; May, 13; June, 2; July, 1; August, 1.

In four more recent determinations of the incubation period, the shortest was between 12.5 and 13 days; another between 13 days and 13 days and 7 hours; one approximately 13 days; and one between 13 and 14 days. These records, along with seven earlier determinations, indicate that in Costa Rica the incubation period of the Blue Tanager is usually from 13 to 14 days.

A Bigamous Family

While watching a Tropical Kingbird's nest on the hillside behind our house on 9 May 1954, I repeatedly saw three Blue Tanagers fly together to a neighboring annonaceous tree. After a little searching I found their nest, 10 feet up in a narrow crotch, against the trunk. With a mirror, I saw four eggs, twice as many as I have found in any other Blue Tanagers' nest in southern Central America. Two of these eggs were rather light blue and the other two much paler, more grayish, but the markings of all four were similar. They had evidently been laid by different birds.

Setting my blind in front of this unusual nest, I watched on the morning of 11 May. I soon learned to distinguish the two tanagers who took turns on the eggs. On one the plumage around the eyes and on the hindhead looked moth-eaten, as though some feathers had been lost or broken. The other had perfect plumage and was much prettier; she closely resembled the male but did not sing, as he did. The female with perfect plumage was clearly dominant over the female with damaged plumage. This was an unmistakable case of bigamy.

During the four hours of my vigil the four eggs were kept constantly covered by the two females, sitting alternately; but they did not take equal shares in incubation. The dominant female, A, sat when she wanted to; the subordinate bird, B, sat when the other permitted her to occupy the nest. The latter always hopped meekly from the nest as A approached; once, when she was slow to leave, she received a peck on the head, which explained why she looked so unkempt. While A incubated B spent much time hopping restlessly close around the nest, or else preening nearby, obviously desiring to cover the eggs but unable to do so. Once B sat on the nest's rim with her breast over A, but continued so only for a minute. The moment that A stretched her wings and left the eggs for an outing, B always went to sit on them, so that they were left uncovered only for an instant.

Female A took seven sessions ranging from 18 to 32 minutes and averaging 25.6 minutes. Her seven recesses varied from four to 10 minutes and averaged 7.6 minutes. She was in the nest for 77.2 per cent of the four hours. Her pattern of incubation was much like that of a Blue Tanager incubating alone (Skutch, 1954:196-197), except for the absence of very short sessions, which was perhaps caused by reluctance to relinquish the eggs to her rival. Female B's sessions coincided with A's recesses; she was in the nest only 22.8 per cent of the time.

The bigamous male fed A on the nest three or four times, but he gave nothing to B while she sat. This bestowal of his favors did not necessarily indicate partiality to A but can be otherwise explained. The male usually accompanied whichever female was absent foraging, then returned with her to the nest tree. Whenever he returned with A, B promptly left the nest before she could be fed there; but when he returned with B, A would continue to occupy the nest and receive whatever food he took to it. Female B, who perforce incubated less than A, spent much more time with the male and may have been fed by him when beyond view.

It was interesting to find the dominant personality insisting upon doing the larger share of the work, leaving the other free to roam. In a corresponding situation among humans, the reverse would probably occur. I regretted that I did not discover this nest before the eggs were laid, to learn whether continuous incubation would accelerate hatching.

By 13 May this nest held four nestlings, all about the same age. A week later only three remained, appearing crowded in a nest that usually shelters only two. All three parents fed them, often coming to the nest together. At times two, standing on opposite sides of the bowl, would deliver food simultaneously, while the third waited until they had finished; or the three might feed the nestlings in quick succession. Because the food was mostly carried inside the parents' mouths and bulging throats, it was difficult to learn its nature. Sometimes after feeding the nestlings the male would alight on a neighboring twig and sing. Because of their rapid movements it was now more difficult to distinguish the two females than it had been while they incubated, to learn whether one brought food more frequently than the other, but it was certain that both fed very often. In the four hours from 05:30 to 09:30, the three parents fed the three eight-day-old nestlings 142 times, or at the rate of 11.8 times per nestling per hour. There was one adult for each nestling, as at normal nests with two young attended by two parents. The nestlings were not brooded.

I noticed no antagonism between the two females, who now worked in perfect amity attending their combined families. They often hunted together over the slender terminal twiglets of surrounding trees, both those with leaves and those that were bare. Frequently they hung head downward to examine the tips of the twigs.

When the nestlings were 11 days old and becoming feathered, I found one on the ground beneath the nest tree. The following afternoon a flightless nestling was again on the ground and only one remained in the nest. On the morning of 27 May two young tanagers were perching in the nest tree and the nest was empty. They were only 14 or 15 days old. Blue Tanagers rarely leave before the age of 17 or 18 days and I do not know what caused these young to abandon their nest prematurely. Two days later the nestlings, now old enough to fly, had vanished.

I cannot explain how this bigamous trio arose among Blue Tanagers that normally maintain monogamous bonds throughout the year. Polygamy has rarely been recorded in the tanager family. Willis (1961) noticed cases of bigamy in the Red-throated Ant-Tanager and Red-crowned Ant-Tanager. One male of the latter had two mates at the same time, but they had separate nests instead of laying their eggs together like the Blue Tanagers. Although Scarlet-rumped Black Tanagers are usually monogamous, some of the more numerous females, unable to find an unmated male, form an irregular attachment with one already mated, a situation disclosed later by the female's failure to receive help in feeding her young, such as regularly mated females of this species receive from their partners. The absence of a male attendant at a nest of the Silver-beaked Tanager in Venezuela suggested a similar situation in this related species (Skutch, 1968). These are the only instances of departure from strict monogamy in the tanager family that have come to my attention.

Summary

This account adds information to an earlier life history of the Blue Tanager.

To a diet consisting largely of fruits and insects, Blue Tanagers add green leaves, including those of the leguminous scrambler *Cassia indecora*. On the island of Tobago they take sugar from dining tables.

They roost amid dense foliage, the members of a pair separated from each other by their own length or more, apparently never in contact.

Regional song dialects have been noticed.

Two instances of anting have been observed, both in trees.

In Venezuela Blue Tanagers sometimes nest in the many-chambered, hanging structures built of interlaced twigs by Rufous-fronted Thornbirds. More often they build in trees and shrubs. They lay two eggs.

In Costa Rica a nest was found at 7,300 feet above sea level.

In the Valley of El General, Costa Rica, eggs are laid from February to August, with the peak in March, April, and May.

A case of bigamy was studied. Two females mated with the same male laid in the same nest, a total of four eggs. The dominant female incubated when she desired to and the subordinate female during the absences of the former. The male fed the dominant female on the nest but not the subordinate female. All three parents fed the nestlings at the rate of 11.8 times per nestling/hour when they were eight days old. The nestlings left the nest prematurely and apparently only two of the four survived.

DUSKY-FACED TANAGER

Mitrospingus cassinii

The Dusky-faced Tanager is a slender bird about 6¾ inches in length. In both sexes the forehead, face, and chin are grayish black. The center of the crown and hindhead are yellowish olive-green. The remaining upper parts and the long tail are dark grayish olive. The throat is gray and the remaining under parts are bright yellowish olive-green, tinged with brown on the under tail coverts. The rather long, slightly downcurved bill has a dark upper mandible and a yellowish lower mandible. The legs and toes are dark. The eyes are dull yellow, a color so rare in the tanager family that it almost serves for the identification of this species. (The only other yellow-eyed tanager that I know is the Magpie-Tanager of South America, whose black and white plumage is distinctive in the family.)

The Dusky-faced Tanager ranges from Costa Rica to western Ecuador. In the former country it is confined to the lowlands and foothills of the wet Caribbean side and hardly ascends to 2,000 feet above sea level. From their home in the densely tangled vegetation along the banks of lowland streams and about the edges of swampy openings in the forest, these tanagers make excursions into neighboring clearings grown up with rank vegetation and into the marginal parts of the forest itself, but they rarely penetrate far into unbroken woodland. In family groups they move restlessly through the thick shrubbery, revealing their presence by their constant chatter of sharp notes, rarely delaying long in one place. It is difficult to detect more than two or three at a time, or to learn the number in a flock, except when they fly one by one across a trail or opening, to vanish immediately amid the leafage on the farther side. Counting them so, I have found up to 10 in a family party, although usually there are no more than six or eight. If these tanagers have notes more melodious than their harsh calls, in two seasons at La Selva, where they were fairly abundant, I failed to hear them. These slender, garrulous, yellow-eyed birds remind one more of jays than of tanagers. When I first met them, in the forest undergrowth near a stream, I hardly knew in what family to place them.

Dusky-faced Tanagers subsist upon a mixed diet of insects, spiders, and small fruits. To find the former they search through the foliage and probe curled dead leaves, sometimes hanging head downward. They eat a variety of berries, especially those of the Melastomaceae, Rubiaceae, and Solanaceae, and likewise the seeds of the scandent grass *Lasiacis,* and the small, hard seeds of the tree *Alchornea costaricensis,* each enclosed in a thin red aril.

Nesting

From my first encounter with these unusual tanagers, I felt sure that they had peculiar habits which would well repay study, and during two seasons at La Selva I tried hard to find their nests. To locate them by searching through tangles where every forward step had to be won by vigorous wielding of the machete seemed an unpromising endeavor, so I chose the alternative procedure of depending upon the birds to guide me to them. But this method was also unproductive, for the restless flocks in which these tanagers traveled even in the breeding season moved through the thickets faster than I could follow. The single nest that I finally discovered was found in peculiar circumstances, which themselves reveal something of the character of these birds. While crossing a bridge over a stream flowing between high banks, I noticed a Dusky-faced Tanager arranging fibrous materials in a shrub amid the tangled waterside vegetation about 100 feet upstream. While this bird worked, several other individuals of its kind flitted noisily through the surrounding bushes. The builder soon left along with the others; and on this and later mornings I passed hours vainly watching for the work to be resumed. The tanagers, however, continued to pass in and out of the thicket, where they were difficult to see, and finally I noticed that they were carrying food. After trying a number of points on the higher opposite bank, I at last found a spot whence I could look down at a nest that already held two well-grown nestlings. I did not then attempt to reach this nest, which would have entailed much disturbance of the surrounding tangle, but continued to watch from across the stream.

This nest, collected after the young had flown, was a bulky open cup slung between two upright branches a few inches apart, with no support below. It was composed almost wholly of long, brown, thread-like, dry, pistillate inflorescences of *Myriocarpa,* some of which had been wound around the supporting branches while many more were spread untidily over surrounding leaves and twigs, especially those below the nest. There was a slight admixture of wiry rootlets and slender rhizomes of an epiphytic fern, with the small, entire leaves attached and living. The sparse lining consisted of black fungal filaments or "vegetable horsehair." The nest measured 5½ by five inches in over-all diameter by 3¼ inches high. Inside it was 3½ by three inches in diameter by two inches deep. This nest was in a young tree, slightly above the densest of the surrounding vegetation and about 10 feet above the water level, which after three nearly rainless weeks had fallen low. But the deposit of silt on the foliage revealed that the creek sometimes rose above the nest site, which at any season it might do in the course of a day, when its current was held back by floodwater in the Río Puerto Viejo, a few hundred yards downstream.

When found on 30 April 1968 this nest held two nestlings, which from the opposite bank I could see whenever they stretched up their heads with gaping mouths, revealing the bright red interior.

On the afternoon of the day the nest was discovered these nestlings were fed only seven times in the three hours from 14:10 to 17:10. They were no longer brooded. On the following day, 1 May, they received 43 meals in the six hours from 05:25 to 11:25. For long intervals the attendants of the nest would remain beyond sight and hearing. Presently we would hear their continuous chatter of sharp notes, growing louder as the flock gradually approached, foraging through the tangled vegetation along the creek. Finally the foliage a few yards from the nest would begin to sway and then a yellow-eyed bird, bearing an insect in its sharp bill, would appear on a slender, exposed twig, nervously flitting its wings both together. It might fly directly across the clear space separating it from the nest or it might vanish into the dense foliage again before approaching the nest. Meanwhile, the two nestlings would stretch up their gaping red mouths in anticipation of a meal. The attendant never alighted on the nest itself but always clung to the upright stem beside it, several inches above the nestlings, and reached downward with inverted head to place the food in an open mouth. Then, sometimes carrying a white dropping in its bill, it would dash back into the concealing thicket from which many sharp notes continued to emerge.

So careful were these tanagers to avoid exposure when near the nest that we rarely had more than two, bringing or delivering food or flying back to concealment, in view at the same time. But it so frequently happened that three fed the nestlings within a minute or so, that we had no doubt that at least three individuals were involved—there was hardly time for one bird to have found another insect and repeated the feeding. Indeed, twice during the morning there were indications that four birds were bringing food. We suspected that the number of attendants was greater than this, for the party that repeatedly passed through the thicket around the nest contained about seven birds; but because we could not distinguish them individually, nor keep them long in view, we could not prove this. Although the tanagers preferred to approach the nest in a flock, even if only one or two bore food, occasionally a lone bird would bring something to the nestlings, then promptly vanish. Some members of the group seemed timid in our presence 50 feet across the creek and partly screened by foliage and flitted around near the nest with food that they never delivered.

Although Dusky-faced Tanagers consume many berries, we recognized only invertebrate animals among the food they gave their nestlings, always one article at a visit. Grasshoppers and other insects, an occasional green caterpillar or a spider, were among the objects that could be identified.

On 4 May we watched the nest from 05:40 to 10:40. The nestlings received 28 meals in the five hours. Now fairly well feathered, they were quite active, moving around much, preening their fresh plumage, flapping their wings on which the remiges were still partly ensheathed. One stood briefly on the nest's rim. On the following day they left. In late May this flock contained 10 birds, which doubtless included the two juveniles from this nest, now well grown.

Although we could find no other nest, on 12 May I watched three of these tanagers carrying food into a dense thicket along a different stream, and concluded that they were taking food to fledged young who lurked invisible amid the lush, low vegetation. Apparently in this social species many nests are attended by three or more birds. The extra individuals seem to be unmated helpers, probably in many instances older siblings of the nestlings they feed.

SUMMARY

Dusky-faced Tanagers are found chiefly in dense vegetation along lowland streams and about the edges of swampy openings in the forest. Throughout the year they live in family groups of up to six or eight, or rarely more, individuals who move restlessly through tangles impenetrable by man. They subsist on a variety of small fruits and of insects, which they find amid living foliage and curled dead leaves.

While foraging they constantly repeat sharp, harsh notes, and they seem to have no song.

In late April a single nest was discovered in a streamside tangle. The two nestlings were fed by at least three, and probably more, of the seven grown tanagers who frequented the vicinity. On each visit the attendant brought a single insect or spider, held conspicuously in its sharp bill, and delivered it while clinging above the nest in an inverted position. In six hours 43 meals were delivered.

The extra attendants appear to be unmated helpers, probably older siblings of the young that they feed.

Family FRINGILLIDAE

BLACK-FACED GROSBEAK

Caryothraustes poliogaster

The Black-faced Grosbeak, also known as the Bishop Grosbeak, is a stout finch from 6½ to seven inches long. In both sexes there is a black patch around the base of the bill, covering the lores, forepart of the cheeks, chin, and upper throat, appearing square when viewed from the front. The rest of the head, neck all around, and breast are yellow, which is brightest on the forehead and lower throat, while posteriorly it is tinged with olive as it blends into the yellowish olive-green of the back. The rump and upper tail coverts, sides and flanks are gray, which fades to grayish white on the center of the abdomen. The wings and tail are olive-green. The short and very thick bill is black, becoming blue-gray at the base; the eyes are brown; and the legs and toes are gray.

This grosbeak ranges through the lowlands and foothills of the eastern side of the continent from southern Veracruz in Mexico to near the Canal Zone in Panama, where alone it crosses to the Pacific watershed. There is an old record from the Canal Zone, but no recent report of its occurrence there (Eisenmann and Loftin, 1967b:32). Altitudinally it extends from near sea level up to about 3,300 feet in the northern part of its range in Mexico (Ridgway, 1901:656), British Honduras (Russell, 1964:176), and Guatemala (Griscom, 1932:353), but it is rarely found above 2,500 feet in Costa Rica. It inhabits the heavy lowland forests where it usually forages high in the trees, but it is more often noticed about their edges and in clearings with scattered tall trees, such as shaded cacao plantations and pastures, where it may hunt nearer the ground, although even here it rarely descends below 20 feet. A social bird at all seasons, the Black-faced Grosbeak wanders restlessly in loose, straggling flocks, the members of which can rarely be seen all at once or accurately counted. Often these flocks contain a dozen or two of loosely associated grosbeaks, mingled with which are frequently a variety of other small arboreal birds. Even while attending nests these finches are gregarious, making no attempt to defend territories, with interesting consequences that we shall presently consider.

Food

Black-faced Grosbeaks are versatile foragers who include large quantities of both fruit and insects in their diet. Among the former I have noticed the fleshy orange fruits of the Mexican rubber tree *Castilla;* the red berries of *Hamelia patens;* the whitish berries of the epiphytic cactus *Rhipsalis Cassutha* and of a scandent species of *Tournefortia;* the blue or blackish berries of melastomaceous shrubs; and the small seeds, enclosed in thin red arils, of the tree *Alchornea costaricensis.* Berries are usually worked between the thick mandibles until the skin is detached and dropped, only the pulp with included seeds being swallowed. With slow, deliberate movements the grosbeaks search for insects amid the foliage of trees. They may hang with head downward to reach caterpillars on lax foliage, fly up to pluck an insect from a higher bough, and occasionally dart outward to catch one in the air. They investigate curled dead leaves and the interior of tank bromeliads. When in May the new foliage of cacao, trees of *Goethalsia meiantha,* and species of *Inga* was being devoured by hordes of green caterpillars, the grosbeaks hunted them assiduously and fed them to their young.

Voice

While foraging the grosbeaks frequently repeat a dull, "thick" note which when the birds are near is perceived to be slightly buzzy: *z-z-z-t.* Mingled with this harsh note is often a low, soft *tweet* that is not so obvious. A longer utterance, frequently heard from wandering flocks as well as nesting pairs, is a scarcely melodious *chip chip chip* . . . , three or four to rarely as many as six of these notes in a series. Not only males but likewise females, sometimes while they incubate, give this refrain, which is doubtfully to be considered their song. This doubt does not apply to the softer, more melodious phrases, of three to six notes, which I did not notice until nesting was about to begin, and usually from solitary or paired birds rather than from those in large flocks. Variants of this song sounded like *weet cher cher weet cher; cher che weet, cher che weet;* and sometimes *glad to meet you.* I never knowingly heard this sweeter song from a female. These appealing verses were heard far more rarely than the prosaic *chip chip chip.* Even when singing persistently away from a flock the grosbeaks commonly alternated between the two types of songs, repeating each a number of times before shifting to the other; and the less musical phrase usually predominated.

Anting

One cloudy afternoon, among the shade trees of a cacao plantation, I watched a grosbeak "ant." It picked something, apparently an ant, from a branch and rubbed it beneath a half-lifted wing, repeating this several times. It did not bend its tail forward beneath its body, as birds engaged in this puzzling activity often do. I could not see whether it finally ate what it held in its bill. Although in the temperate zones even arboreal birds frequently ant on the ground, the many tropical species that I have watched apply ants to their feathers have almost invariably done so in trees or shrubs.

Nest Building

Despite the abundance of Black-faced Grosbeaks at La Selva, we discovered only one nest in the two seasons we passed there. Apparently the nests are tucked away amid the epiphytes that grow so profusely on the trees, where it is scarcely possible to find them unless the birds themselves reveal their location by their visits. Such, at least, was the situation of the single nest that we saw. This was in a small, sickly tree, profusely covered with aroids, bromeliads, ferns, orchids, gesneriads, mosses, and other epiphytes in the narrow clearing behind the house, bordered on three sides by forest trees. On 9 May 1967 a grosbeak sat for several minutes in a niche amid these epiphytes, while its mate perched nearby and sang. On 14 May, a sunny day after a rainy week, a grosbeak again sat in this cranny while its mate sang. Then both members of the pair plucked dying leaves from neighboring trees and tried to carry them to the site; but a pair of Black Variable Seedeaters nesting in the same tree protested vehemently; and soon the grosbeaks flew away, abandoning the attempt to build here.

Late on the morning of 9 April of the following year, we noticed a pair of grosbeaks searching for a nest site in the same small tree between the house and the forest. They hopped in and out among the epiphytes while one sang *chip chip chip chip* and both uttered dull, buzzy notes. Presently they flew across to the forest's edge and started to gather material. After bringing a few pieces, they flew away, to return after a brief interval and bring more. Again and again during the late forenoon and early afternoon they reappeared, each time staying but a short while and bringing at most a few pieces. Most of their material was gathered from the epiphytes on neighboring forest trees. Although both sexes built, one, doubtless the female, seemed to do the major share.

The nest grew very slowly, for there was no sustained building. At long intervals the pair would appear, go to the nest a few times, then fly beyond sight and hearing. The male never sang persistently from an exposed perch, as many finches do, to proclaim his possession of territory and his determination to defend it; he was present only while the pair built. The nest remained such a slight structure that I could not decide whether it was finished until the first egg was laid 12 days after building began. It was then a shallow bowl about three inches in internal diameter by 1¼ inches deep. The loose outer layer consisted of slender, creeping rhizomes of an epiphytic fern, up to two feet in length, with the thick entire leaves, an inch or less long, still attached and green. Pieces of dead dicotyledonous leaves made the thin middle layer. The lining was unlike that of any other nest I have ever seen. It consisted of the leaves of an epiphytic bromeliad, apparently a species of *Tillandsia,* up to 11 inches long, tapering from a base rarely more than 1/16 inch wide to a hair-thin apex.

These were gathered and coiled down into the bowl while fresh and soft, and many remained green until after the nestlings flew. All the components of the nest were evidently collected in trees rather than from the ground. The slight, flimsy structure was hidden in a cranny amid the air plants, beneath a big tank bromeliad and a large-leafed aroid that shielded it from rain. It was only nine feet above the ground.

Eggs

Two eggs were laid between 20 and 22 April. The third and last was deposited before 10:15 on 23 April. These eggs were dull white, mottled and spotted all over, very heavily on the thicker end, with bright shades of brown. In shape they were long-ovate and measured 26.3 by 17.2, 26.0 by 16.9, and 26.5 by 16.8 mm.

After the eggs were laid I encircled the slender trunk of the nest tree with a wide metal band, to prevent animals from climbing up to them.

Incubation

Although the sexes could not be distinguished by appearance, nor with certainty by voice, our failure to witness a changeover in more than 18 hours of watching led us to conclude that only the female incubated. On the clear day of 26 April we watched from 05:10 until 18:00. The female grosbeak, who had been sitting silently since daybreak, first left her eggs at 05:54 and she settled on the nest for the night at 17:10, nearly an hour before it was dark. In an active day of 11¼ hours, she took seven sessions on the nest, ranging from 37 to 123 minutes and averaging 60.4 minutes. Her longest session, lasting more than two hours, was taken between 14:31 and 16:34 in the afternoon. Her eight recesses ranged from 22 to 40 minutes and averaged 31.6 minutes. She covered her eggs for 62.6 per cent of her active period.

After an absence from the nest, during which the pair of grosbeaks foraged beyond sight and hearing, the two nearly always returned together, singing the *chip* song and repeating the *z-z-z-t* call as they made their way through the crowns of the trees at the forest's edge to a point near the nest. Then they would fly sharply downward toward the little nest tree, close together, as though racing each other to this goal. The female would alight in the nest tree while the male swept past it and rose to the tall trees at the opposite side of the narrow clearing, describing a deep catenary loop between opposite edges of the forest. After alighting in the nest tree the female might give the *chip* song, interspersed with the thick buzzy call, before she settled on her eggs; and sometimes she would continue these utterances, in a voice as loud as the male's, even while she sat. The male then flew out of sight but sometimes he would return to sing in neighboring trees while she incubated, usually with *chip*'s, more rarely with the more melodious song.

Often he was present when she ended her session and they flew away together, across the river or toward a neighboring cacao plantation.

The spectacular fashion in which the male escorted his mate to the nest, as though trying to get there first, reminded me of the similar habit of certain birds that build closed nests, such as euphonias and Black-fronted Tody-Flycatchers. When birds with open nests accompany their mates back to their eggs, they nearly always do so in a less dashing fashion. Although the grosbeaks had a shallow open nest, epiphytes clustered so thickly around that it was almost enclosed.

We watched this nest again through the morning of 3 May while rain fell steadily. The female now incubated more constantly than on the clear morning, taking four sessions that lasted 43, 103, 61, and 75 minutes, and three recesses of 9, 18, and 38 minutes duration.

On 26 April, when we watched all day, there was no indication that more than one male and one female were interested in the nest, although once two grosbeaks were heard singing nearby while the female incubated. On the afternoon of 29 April three grosbeaks came to the nest tree and on the following day four were seen near the nest. We noticed no antagonism between them. On the rainy morning of 3 May, however, two grosbeaks were regularly escorting the female to the nest, exactly as one had done on previous days. After maneuvering into position in a tree at the forest's edge, the three darted down toward the nest close together, as though moved by a single impulse. Then, after the female alighted in the nest tree, the other two would continue together up to the edge of the forest. These two seemed to keep close company, and while the female incubated they would often sing together in neighboring trees. Sometimes she sang in the nest, answering them. We did not learn the sex of the third bird, who thenceforth became a constant companion of the mated pair.

On 26 April, ten minutes after the female settled on the nest for the night, her mate approached by slow degrees and seemed to give her something, but food was not clearly seen. Early on the morning of 3 May one of the two birds who then accompanied the female went to the nest tree with something green in its bill, but hesitated to take it to the nest. It flew off and returned to the nest tree bearing the same object, whereupon the female flew away and the other followed, still holding the food. I could not see whether she finally received it. The following afternoon, after the two companions had escorted the female to the nest and returned to the edge of the forest, one plucked a fruit and passed it to the other. I could not tell whether the female's mate fed the third bird, or the reverse. One morning in early May, while I watched grosbeaks eating caterpillars amid young foliage, one presented a caterpillar to another, who might have been its mate. But it is not impossible that the sociable grosbeaks, like Cedar Waxwings and certain other highly gregarious birds, pass food to adult companions other than their mates.

The last egg had been laid before 10:15 on 23 April and the last nestling hatched between 12:30 and 16:00 on 6 May. The incubation period was accordingly between 13 and 13½ days.

Nestlings

At 06:25 on 6 May there was a single nestling which had hatched since the preceding afternoon. It had pink skin, shaded by sparse but rather long gray down which formed a fringe or halo around its bald crown. The interior of its mouth was deep red, the corners yellow. The empty shell had already been removed.

Feeding.—At 07:05 on 6 May I began to watch continuously. The female was absent but after a few minutes she returned with the other two grosbeaks. She seemed to feed a nestling while standing on the nest's rim, ate another empty shell, then settled in the nest. After brooding for nine minutes she left, joined her two companions, and all flew away. In two minutes she returned with one other grosbeak. Nothing was visible in their bills but after the female entered the nest the companion gave food to her and she passed it to the nestlings. She brooded for nearly half an hour, then left as both companions arrived with food in their bills. For about five minutes one of the latter, uttering soft notes, tried to feed the nestlings while the female and the other companion hopped around the nest, the latter still with a full bill. The first feeder seemed at last to succeed in delivering what it had brought, then flew off with the female. The second companion then went to the nest, but instead of trying to feed the nestlings it hurried after the other two, still holding the food. By the middle of the morning all three were feeding the nestlings, as they continued regularly to do until the young birds flew away.

We spent the whole of 8 May watching the nest. The three nestlings, now two days old, were fed 50 times by their three attendants, or at the rate of 1.4 times per capita per hour for the active period between the female's first morning departure from the nest at 05:25 and her return for the night at 17:44. Often all three attendants came together, announcing their approach with the *chip* song and harsh calls. When they had reached a point at the forest's edge near the nest tree, the two companions would escort the female to it in the same dashing fashion as while she incubated. Leaving her there to feed and perhaps also brood the nestlings, they swept past and rose again into the trees beside the clearing. Then, if they had brought food, they would descend to the nest to deliver it, sometimes giving it to the brooding female, who rose up and passed it to the nestlings beneath her, sometimes feeding the nestlings directly. If the female stayed to brood the other two would fly away together, and perhaps return together with more food before she left. Throughout the nestling period food for the young was not regurgitated but carried in the attendants' mouths and bills, which were usually slightly parted when they arrived with a meal. But so little of it was visible between the thick mandibles that it was difficult to learn its nature. Often it seemed to be fruit, rarely it was clearly an insect. The droppings of the two-day-old nestlings were nearly always swallowed by the adults, after they had delivered their food.

On 14 May the three nestlings, now eight days old, were fed 93 times in the course of a rainy day. The first meal was delivered at 05:26, the last at 17:50, and the nestlings were fed at the rate of 2.5 meals per capita per hour. The nestlings were now no longer brooded and all three attendants usually came and left together. Arriving from afar with food, the grosbeaks, singing and calling, first alighted in the trees at the edge of the forest, some distance from the nest tree. Then, by short flights from branch to branch, they made their way along the sides of the clearing until opposite the nest. Then they flew down, usually one by one, to the little nest tree. Nearly always they fed in succession, exceptionally simultaneously from opposite sides of the nest. Usually each attendant fed two or three nestlings at a visit. After delivering the food it might take a dropping, which it either swallowed or carried away in its bill. Sometimes the first to feed would wait for the other two, and all fly away together; but at other times they left one after another, all going in the same direction. Rarely one of the trio would come without food, doubtless because it had found nothing by the time the other two were ready to proceed to the nest. Once, in order not to be left behind, one carried away the food it had brought, hurrying after its companions.

Helpers.—Although only these three were regular attendants, from time to time additional grosbeaks visited the nest tree. On the afternoon of 8 May the three attendants, coming with food, were followed by two more birds who did not feed. There was much singing near the nest, then all five flew away. A little later, however, four, including the mother, came and fed the nestlings. On a number of later occasions four or five grosbeaks gathered around the nest, and once again four appeared to feed the nestlings; but, unfortunately, this time I did not have my binoculars to make sure of this. One day seven grosbeaks were in the nest tree flitting around and singing. Usually when extra birds arrived no enmity was noticed; but once, when four were present, one chased another mildly, not out of the nest tree, which had only a few small branches. We never noticed any antagonism between the three regular attendants.

These wandering grosbeaks who accompanied the parents to the nest, perhaps fed a nestling, then passed on, suggest how the third attendant became attached to it. Since, despite much watching, both intensive and occasional, of this nest in front of our windows, the third bird was not noticed until after incubation was well started, evidently we have here an instance of an unmated helper, of unknown sex, rather than of polyandry or polygyny. Although, to my great regret, we could find only one grosbeak's nest, I suspect that such helpers occur not infrequently in this species; they are in keeping with its character which manifests extreme sociability, lack of territorial defense, and interest in neighbors' nests. Intraspecific helpers are rather exceptional among the Fringillidae, doubtless because they are, on the whole, strongly territorial birds; they are of more regular occurrence among the tanagers, in which territorialism is less developed (Skutch, 1961b). Most of the cases of "helpfulness" that have been recorded for the Fringillidae have been of interspecific helpers, which are always more or less accidental rather than of regular or frequent occurrence, as is true of many intraspecific helpers. In a number of cases where three finches of the same species have been found attending the same brood, polygamy has been demonstrated or suspected, as in the Variable Seedeater (Gross, 1952) and the House Finch (Robert S. Woods *in* Bent, *et al.*, 1968:295, 299).

Brooding.—The young grosbeaks were brooded surprisingly little and only by their mother, as far as we could tell. Even on 8 May, when they were naked mites two days old, they were brooded only 27.5 per cent of the day. Most of this brooding was done in the morning, when once the parent covered the nest for 51 minutes continuously. In the afternoon she brooded only twice, for eight and then 29 minutes, leaving the tiny nestlings exposed for intervals of 86, 90, and 128 minutes. Although nights were often rainy, nocturnal brooding ceased when the nestlings were only six days old. On the rainy day of 14 May, when the developing plumage of the eight-day-old nestlings was still ensheathed and gave little protection, they were not once brooded. In ceasing to brood before the young were feathered, the grosbeaks behaved like certain birds that rear their families in holes or covered nests, such as motmots and euphonias. Doubtless they depended on the epiphytes that roofed over the nest to keep it dry. In contrast to these grosbeaks, a female Scarlet-rumped Black Tanager, who raised two young in a neighboring open nest, covered them nightly until they left at the age of 11 days. Variable Seedeaters brood their young in open nests by night until they are 11 or 12 days old, well feathered, and ready to fly.

Departure.—By the morning of 17 May, when 11 days old, the three young grosbeaks were fairly well covered with feathers and their remiges were expanding, but their heads were still nearly naked. They preened and flapped their wings in the nest. At 16:35 in the afternoon of the same day, after a long-continued torrential rain, one of the young birds flew from the nest in the presence of its attendants. It covered about 40 feet on a slightly ascending course, to alight on an exposed twig at the forest's edge where it was promptly fed. It remained through the night on the perch where it first alighted. At dawn my attention was drawn, by the scolding of the adult grosbeaks and a variety of other birds, to a six-foot boa that was coiled up into a ball on a branch only 10 yards from where the fledgling roosted.

Leaving the big snake to rest inertly in the same spot all day, the grosbeaks divided their attention between the fledgling at the forest's edge and the two nestlings still in the nest. At 07:01 another left, flying about 35 feet on its first attempt. It alighted clinging below a slender twig at the edge of the forest but promptly pulled itself up to perch competently. At the moment when it first took wing the attendants were present, flying around and calling excitedly; but they did not appear to coax the fledgling from the nest. By 10:30 that morning the third young grosbeak had flown and the whole family had passed from view into the leafy mazes of the neighboring forest. We did not again see the fledglings in or beside the clearing where they had hatched.

These young grosbeaks had flown from the nest when 11½ to 12 days old. At this age they were yellowish brown on the hindhead, nape, and breast. Their back and primaries were olive, secondaries more greenish olive, and abdomen gray. There was still no black on their incompletely feathered heads.

Older juveniles, still begging for food, travel with flocks of adults, but it is now impossible to tell how many feed them.

Summary

The Black-faced Grosbeak inhabits rainforest and adjoining clearings with scattered trees. Social at all seasons, it wanders through the treetops in straggling flocks that may comprise several dozen individuals.

Despite the abundance of these grosbeaks at La Selva, only one nest was found. At a height of nine feet, the shallow bowl was excellently concealed amid epiphytes that burdened a small tree in a narrow clearing.

The first of three eggs was laid 12 days after building began. Only the female incubated. As she returned to her eggs she was escorted to her nest in spectacular fashion by her mate, who flew close beside her as though racing to reach it first.

There was little indication of territoriality. Wandering grosbeaks who alighted close to the nest were not chased away by the parents but at most mildly threatened. After incubation had continued for some days, one of these vagrant birds, of unknown sex, attached itself to the mated pair and was thereafter their constant companion. Now two grosbeaks, indistinguishable from each other, raced the female to the nest, then flew away together while she settled on the eggs. Rarely one brought food to the nest where the female incubated, and once one of her two companions fed the other.

The incubation period was 13 to 13½ days. The nestlings were brooded little, apparently only by their mother, who ceased to cover them by night when they were only six days old and still unfeathered. The epiphytes above them afforded some protection from the frequent nocturnal rains.

Throughout the nestling period the young were regularly fed by three adults, the parents and their helper, who, after brooding ceased, nearly always came and left together. Rarely some other grosbeak passing by noticed the nestlings and fed them. When two days old the three nestlings were fed 50 times by their three attendants during a day of 12⅓ hours. At eight days they were fed 93 times in the course of a rainy day.

When 11½ to 12 days old the young left the nest and promptly entered the neighboring forest.

LITERATURE CITED

Andrew, R. J. 1961. The displays given by passerines in courtship and reproductive fighting: a review. Ibis, **103a:** 315-348; 549-579.

Arbib, R. S., Jr., and F. W. Loetscher, Jr. 1935. Some notes from the Panama Canal Zone—summer 1934. Auk, **52:** 325-328.

Audubon, J. J. 1831. Ornithological Biography, I. Edinburgh.

Belcher, C., and G. D. Smooker. 1936. Birds of the colony of Trinidad and Tobago. Ibis, ser. 13, **6:** 1-35; 792-813.

Bent, A. C. 1940. Life histories of North American cuckoos, goatsuckers, hummingbirds and their allies. Bull. U. S. Nat'l Mus., **176:** 1-506.

Bent, A. C., *et al.* 1968. Life histories of North American cardinals, grosbeaks, buntings, towhees, finches, sparrows, and allies. Bull. U. S. Nat'l Mus., **237:** 1-1889.

Bourliere, F. 1946. Longévité moyenne et longévité maximum chez les vertébrés. Ann. Biol., **22:** 249-270.

Brauner, J. 1953. Observations on the behavior of a captive Poor-will. Condor, **55:** 68-74.

Carriker, M. A., Jr. 1910. An annotated list of the birds of Costa Rica, including Cocos Island. Ann. Carnegie Mus., **6:** 314-915.

Chapman, F. M. 1928. The nesting habits of Wagler's Oropendola (*Zarhynchus wagleri*) on Barro Colorado Island. Bull. Amer. Mus. Nat. Hist., **58:** 123-166.

Chapman, F.M. 1929. My tropical air castle. D. Appleton & Co., New York.

Cherrie, G. K. 1916. A contribution to the ornithology of the Orinoco region. Sci. Bull. Mus. Brooklyn Inst. Arts Sci., **2:** 133a-374.

Conway, W. G. 1967. A door to the out-of-doors. Animal Kingdom, **70:** 2-11.

Davis, T. A. W. 1958. The displays and nests of three forest hummingbirds of British Guiana. Ibis, **100:** 31-39.

Eisenmann, E. 1952. Annotated list of birds of Barro Colorado Island, Panama Canal Zone. Smithsonian Misc. Coll., **117**(5): 1-62.

Eisenmann, E., and H. Loftin. 1967a. Birds of the western Chiriquí highlands, Panama—1967. Florida Audubon Soc., 34 p.

Eisenmann, E., and H. Loftin. 1967b. Field checklist of the birds of the Panama Canal Zone area. Florida Audubon Soc., 35 p.

Ferguson, D. E. 1967. A possible case of egg transport by a Chuck-will's-widow. Wilson Bull., **79:** 452-453.

Friedmann, H., L. Griscom, and R. T. Moore. 1950. Distributional check-list of the birds of Mexico. Part I. Pacific Coast Avif. (Cooper Orn. Soc.), no. 29, 202 p.

Friedmann, H., and F. D. Smith, Jr. 1955. A further contribution to the ornithology of northeastern Venezuela. Proc. U. S. Nat'l Mus., **104:** 463-524.

GANIER, A. F. 1964. The alleged transportation of its eggs or young by the Chuck-will's-widow. Wilson Bull., **76:** 19-27.

GOODALL, J. D., A. W. JOHNSON, AND R. A. PHILIPPI B. 1957. Las aves de Chile. Platt Establecimiento Gráficos, Buenos Aires.

GOODWIN, D. 1962. Notes on the plumages of the Avadavat *Amandava amandava*. Ibis. **104:** 564-566.

GRISCOM, L. 1932. The distribution of bird-life in Guatemala. Bull. Amer. Mus. Nat. Hist., **64:** 1-439.

GROSS, A. O. 1930. Rainbows on wings. Nature Mag., **15:** 249-250.

GROSS, A. O. 1952. Nesting of Hicks' Seedeater at Barro Colorado Island, Canal Zone. Auk, **69:** 433-446.

HAMILTON, W. J., III. 1965. Sun-oriented display of the Anna's Hummingbird. Wilson Bull., **77:** 38-44.

HARTMAN, F. A. 1957. Some additions to nesting data on Panamanian birds. Condor, **59:** 269-271.

HAVERSCHMIDT, F. 1957. The nests of *Pitangus lictor* and *Coryphotriccus parvus*. Auk, **74:** 240-242.

HAVERSCHMIDT, F. 1968. A nest of the Rufous-breasted Hermit with three eggs. Auk, **85:** 693-694.

HUBER, W. 1932. Birds collected in northeastern Nicaragua in 1922. Proc. Acad. Nat. Sci. Phila., **84:** 205-249.

HUNDLEY, M. H., AND C. R. MASON. 1965. Birds develop a taste for sugar. Wilson Bull., **77:** 408.

LAND, H. C. 1962. A collection of birds from the Sierra de las Minas, Guatemala. Wilson Bull., **74:** 267-283.

LAND, H. C. 1970. Birds of Guatemala. Livingston Publishing Co., Wynnewood, Penna.

LAUGHLIN, R. M. 1952. A nesting of the Double-toothed Kite in Panama. Condor, **54:** 137-139.

MILLER, A. H. 1963. Seasonal activity and ecology of an American equatorial cloud forest. Univ. California Publ. Zool., **66:** 1-78.

MONROE, B. L., JR. 1968. A distributional survey of the birds of Honduras. Amer. Ornith. Union, Ornith. Monogr. no. 7, 458 p.

MOORE, R. T. 1947. Habits of male hummingbirds near their nests. Wilson Bull., **59:** 21-25.

MOREAU, R. E. 1944. The Half-collared Kingfisher (*Alcedo semitorquata)* at the nest. Ostrich, **15:** 161-177.

NIETHAMMER, G. 1953. Zur Vogelwelt Boliviens. Bonner Zool. Beiträge, **4:** 195-303.

NOVAES, F. C., AND C. T. DE CARVALHO. 1957. Observações sõbre a nidificação de *Glaucis hirsuta* (Gmelin)—Trochilidae, Aves. Bol. Mus. Paraense Emilio Goeldi, N. S. Zoologia, **1:** 1-11.

PHELPS, W. H., AND W. H. PHELPS, JR. 1958. Lista de las aves de Venezuela con su distribución. Tomo II. Parte I. No Passeriformes. Bol. Soc. Venezolana Ciencias Nat., **19:** 1-317.

Phelps, W. H., and W. H. Phelps, Jr. 1963. Lista de las aves de Venezuela con su distribución. Tomo I. Parte II. Passeriformes. Segunda edición. Bol. Soc. Venezolana Ciencias Nat., **24:** 1-479.

Richmond, C. W. 1893. Notes on a collection of birds from eastern Nicaragua and the Río Frio, Costa Rica, with a description of a supposed new trogon. Proc. U. S. Nat'l Mus., **16:** 479-532.

Ridgway, R. 1901. The birds of North and Middle America. Bull U. S. Nat'l Mus., **50(1):** 1-715.

Ridgway, R. 1911. The birds of North and Middle America. Bull. U. S. Nat'l Mus., **50(5):** 1-859.

Ridgway, R. 1914. The birds of North and Middle America. Bull. U. S. Nat'l Mus., **50(6):** 1-882.

Rowley, J. S. 1966. Breeding records of birds of the Sierra Madre del Sur. Oaxaca, Mexico. Proc. Western Foundation Vertebrate Zool., **1(3):** 107-204.

Ruschi, A. 1965. [Observations on the nesting, incubation and care of young in *Colibri coruscans coruscans* (Gould), done only by the female (Trochilidae—Aves)]. Bol. Mus. Prof. Mello-Leitão, no. 45: 1-9. Abstract in Auk, **84:** 154, 1967.

Russell, S. M. 1964. A distributional study of the birds of British Honduras. Amer. Ornith. Union, Ornith. Monogr. no. 1, 195 p.

Schäfer, E. 1954. Sobre la biología de *Colibri coruscans*. Bol. Soc. Venezolana Ciencias Nat., **15:** 153-162.

Schäfer, E., and W. H. Phelps. 1954. Aves del Parque Nacional "Henri Pittier" (Rancho Grande) y sus funciones ecológicas. Bol. Soc. Venezolana Ciencias Nat., **16:** 3-167.

Skutch, A. F. 1931. The life history of Rieffer's Hummingbird (*Amazilia tzacatl tzacatl*) in Panama and Honduras. Auk, **48:** 481-500.

Skutch, A. F. 1942. Life history of the Mexican Trogon. Auk, **59:** 341-363.

Skutch, A. F. 1944. Life history of the Quetzal. Condor, **46:** 213-235.

Skutch, A. F. 1945. Life history of the Blue-throated Green Motmot. Auk, **62:** 489-517.

Skutch, A. F. 1948. Life history of the Citreoline Trogon. Condor, **50:** 137-147.

Skutch, A. F. 1951. Life history of Longuemare's Hermit Hummingbird. Ibis, **93:** 180-195.

Skutch, A. F. 1953. How the male bird discovers the nestlings. Ibis, **95:** 1-37; 505-542.

Skutch, A. F. 1954. Life histories of Central American birds. I. Pacific Coast Avif. (Cooper Orn. Soc.), no. 31, 448 p.

Skutch, A. F. 1957a. The incubation patterns of birds. Ibis, **99:** 69-93.

Skutch, A. F. 1957b. Life history of the Amazon Kingfisher. Condor, **59:** 217-229.

SKUTCH, A. F. 1958a. Roosting and nesting of araçari toucans. Condor, **60:** 201-219.

SKUTCH, A. F. 1958b. Life history of the White-whiskered Softwing *Malacoptila panamensis*. Ibis, **100:** 209-231.

SKUTCH, A. F. 1959. Life history of the Black-throated Trogon. Wilson Bull., **71:** 5-18.

SKUTCH, A. F. 1960a. Life histories of Central American birds. II. Pacific Coast Avif. (Cooper Orn. Soc.), no. 34, 593 p.

SKUTCH, A. F. 1960b. A forest view of kinkajous. Animal Kingdom, **63:** 25-28.

SKUTCH, A. F. 1961a. Life history of the White-crested Coquette Hummingbird. Wilson Bull., **73:** 5-10.

SKUTCH, A. F. 1961b. Helpers among birds. Condor, **63:** 198-226.

SKUTCH, A. F. 1962a. Life histories of honeycreepers. Condor, **64:** 92-116.

SKUTCH, A. F. 1962b. The constancy of incubation. Wilson Bull., **74:** 115-152.

SKUTCH, A. F. 1962c. Life history of the White-tailed Trogon *Trogon viridis*. Ibis, **104:** 301-313.

SKUTCH, A. F. 1964a. Life histories of hermit hummingbirds. Auk, **81:** 5-25.

SKUTCH, A. F. 1964b. Life history of the Scaly-breasted Hummingbird. Condor, **66:** 186-198.

SKUTCH, A. F. 1964c. Life history of the Blue-diademed Motmot *Momotus momota*. Ibis, **106:** 321-332.

SKUTCH, A. F. 1968. The nesting of some Venezuelan birds. Condor, **70:** 66-82.

SKUTCH, A. F. 1969. A study of the Rufous-fronted Thornbird and associated birds. Wilson Bull., **81:** 5-43; 123-139.

SKUTCH, A. F. 1971a. Life history of the Keel-billed Toucan, Auk, **88:** 381-396.

SKUTCH, A. F. 1971b. A naturalist in Costa Rica. Univ. of Florida Press, Gainesville.

SLUD, P. 1960. The birds of finca "La Selva," Costa Rica: a tropical wet forest locality. Bull. Amer. Mus. Nat. Hist., **121:** 55-148.

SLUD, P. 1964. The birds of Costa Rica: distribution and ecology. Bull. Amer. Mus. Nat. Hist., **128:** 1-430.

SNOW, D. W. 1968. The singing assemblies of Little Hermits. Living Bird, **7:** 47-55.

SNOW, B. K., AND D. W. SNOW. 1971. The feeding ecology of tanagers and honeycreepers in Trinidad. Auk, **88:** 291-322.

STOTT, K., JR., AND C. J. SELSOR. 1961. Association of trogons and monkeys on Barro Colorado. Condor, **63:** 508.

STRESEMANN, E. 1963. The nomenclature of plumages and molts. Auk, **80:** 1-8.

THORPE, W. H. 1958. The learning of song patterns by birds, with especial reference to the song of the Chaffinch *Fringilla coelebs.* Ibis, **100:** 535-570.

TODD, W. E. C., AND M. A. CARRIKER, JR. 1922. The birds of the Santa Marta region of Colombia: a study in altitudinal distribution. Ann. Carnegie Mus., **14:** 1-611.

VAN TYNE, J. 1929. Life history of the toucan *Ramphastos brevicarinatus.* Misc. Publ. Univ. Mich. Mus. Zool., no. 19, 43 p.

WAGNER, H. O. 1954. Versuch einer Analyse der Kolibribalz. Zeit. für Tierpsychologie, **11:** 182-212.

WELLER, M. W. 1958. Observations of the incubation behavior of a Common Nighthawk. Auk, **75:** 48-59.

WETMORE, A. 1968. The birds of the Republic of Panamá. Part 2. Smithsonian Misc. Coll., **150**(2), 605 p.

WILEY, R. H. 1971. Song groups in a singing assembly of Little Hermits. Condor, **73:** 28-35.

WILLIS, E. 1961. A study of nesting ant-tanagers in British Honduras. Condor, **63:** 479-503.

WOLF, L. L. 1969. Female territoriality in a tropical hummingbird. Auk, **86:** 490-504.

WORTH, C. B. 1939. Nesting of some Panamanian birds. Auk, **56:** 306-310.

WORTH, C. B. 1942. Notes on hummingbirds at Chiriquí, Panama. Auk, **59:** 364-368.

INDEX

COMMON NAMES OF BIRDS

SCIENTIFIC NAMES OF BIRDS